Metro Ethernet

Sam Halabi

Cisco Press

800 East 96th Street, 3rd Floor
Indianapolis, IN 46240 USA

Metro Ethernet

Sam Halabi

Published by:
Cisco Press
800 East 96th Street, 3rd Floor
Indianapolis, IN 46240 USA

Printed in the United States of America 1 2 3 4 5 6 7 8 9 0

Library of Congress Cataloging-in-Publication Number: 2002103527

ISBN: 1-58705-096-X

First Printing September 2003

Warning and Disclaimer

This book is designed to provide information about Metro Ethernet. Every effort has been made to make this book as complete and as accurate as possible, but no warranty or fitness is implied.

The information is provided on an "as is" basis. The author, Cisco Press, and Cisco Systems, Inc., shall have neither liability nor responsibility to any person or entity with respect to any loss or damages arising from the information contained in this book or from the use of the discs or programs that may accompany it.

The opinions expressed in this book belong to the author and are not necessarily those of Cisco Systems, Inc.

Trademark Acknowledgments

All terms mentioned in this book that are known to be trademarks or service marks have been appropriately capitalized. Cisco Press or Cisco Systems, Inc. cannot attest to the accuracy of this information. Use of a term in this book should not be regarded as affecting the validity of any trademark or service mark.

Feedback Information

At Cisco Press, our goal is to create in-depth technical books of the highest quality and value. Each book is crafted with care and precision, undergoing rigorous development that involves the unique expertise of members from the professional technical community.

Readers' feedback is a natural continuation of this process. If you have any comments regarding how we could improve the quality of this book, or otherwise alter it to better suit your needs, you can contact us through e-mail at feedback@ciscopress.com. Please make sure to include the book title and ISBN in your message.

We greatly appreciate your assistance.

Publisher	John Wait
Editor-in-Chief	John Kane
Cisco Representative	Anthony Wolfenden
Cisco Press Program Manager	Sonia Torres Chavez
Manager, Marketing Communications, Cisco Systems	Scott Miller
Cisco Marketing Program Manager	Edie Quiroz
Production Manager	Patrick Kanouse
Development Editor	Dayna Isley
Copy Editor	Bill McManus
Technical Editors	Mike Bernico, Mark Gallo, Giles Heron, Irwin Lazar
Team Coordinator	Tammi Ross
Cover Designer	Louisa Adair
Composition	Interactive Composition Corporation
Proofreader	Gayle Johnson
Indexer	Larry Sweazy

Corporate Headquarters
Cisco Systems, Inc.
170 West Tasman Drive
San Jose, CA 95134-1706
USA
www.cisco.com
Tel: 408 526-4000
800 553-NETS (6387)
Fax: 408 526-4100

European Headquarters
Cisco Systems International BV
Haarlerbergpark
Haarlerbergweg 13-19
1101 CH Amsterdam
The Netherlands
www-europe.cisco.com
Tel: 31 0 20 357 1000
Fax: 31 0 20 357 1100

Americas Headquarters
Cisco Systems, Inc.
170 West Tasman Drive
San Jose, CA 95134-1706
USA
www.cisco.com
Tel: 408 526-7660
Fax: 408 527-0883

Asia Pacific Headquarters
Cisco Systems, Inc.
Capital Tower
168 Robinson Road
#22-01 to #29-01
Singapore 068912
www.cisco.com
Tel: +65 6317 7777
Fax: +65 6317 7799

Cisco Systems has more than 200 offices in the following countries and regions. Addresses, phone numbers, and fax numbers are listed on the **Cisco.com Web site at www.cisco.com/go/offices.**

Argentina • Australia • Austria • Belgium • Brazil • Bulgaria • Canada • Chile • China PRC • Colombia • Costa Rica • Croatia • Czech Republic Denmark • Dubai, UAE • Finland • France • Germany • Greece • Hong Kong SAR • Hungary • India • Indonesia • Ireland • Israel • Italy Japan • Korea • Luxembourg • Malaysia • Mexico • The Netherlands • New Zealand • Norway • Peru • Philippines • Poland • Portugal Puerto Rico • Romania • Russia • Saudi Arabia • Scotland • Singapore • Slovakia • Slovenia • South Africa • Spain • Sweden Switzerland • Taiwan • Thailand • Turkey • Ukraine • United Kingdom • United States • Venezuela • Vietnam • Zimbabwe

Printed in the USA

About the Author

Mr. Halabi is a seasoned executive and an industry veteran with more than 18 years of experience marketing and selling to the worldwide Enterprise and Carrier networking markets. While at Cisco, he wrote the first Cisco Internet routing book, *Internet Routing Architectures*, a best-seller in the U.S. and international markets. He has held multiple executive management positions in the field of marketing, sales, and business development and has been instrumental in evolving fast-growing businesses for the Enterprise and Carrier Ethernet markets.

About the Technical Reviewers

Mike Bernico is a senior networking engineer at the Illinois Century Network. In this position, he focuses primarily on network design and integrating advanced network services such as QoS, IP Multicast, IPv6, and MPLS into the network. He has also authored open-source software related to his interests in new networking technologies. He enjoys reading and spending time in the lab increasing his knowledge of the networking industry. He lives in Illinois with his wife Jayme. He can be contacted at mike@bernico.net.

Mark Gallo is a technical manager with America Online. His network certifications include Cisco CCNP and Cisco CCDP. He has led several engineering groups responsible for designing and implementing enterprise LANs and international IP networks. He has a BS in electrical engineering from the University of Pittsburgh. He resides in northern Virginia with his wife, Betsy, and son, Paul.

Giles Heron is the principal network architect for PacketExchange, a next-generation carrier providing Ethernet services on a global basis. He designed PacketExchange's MPLS network and has been instrumental in the development of its service portfolio. A cofounder of PacketExchange, he previously worked in the Network Architecture group at Level(3) Communications. He is coauthor of the draft-martini specification for transport of Layer 2 protocols over IP and MPLS networks and the draft-lasserre-vkompella specification for emulation of multipoint Ethernet LAN segments over MPLS, as well as various other Internet drafts.

Irwin Lazar is practice manager for Burton Group in its Networks and Telecom group, managing a team of consultants who advise large end-user organizations on topics including network architecture and emerging network technologies. He administers The MPLS Resource Center (http://www.mplsrc.com) and is the conference director for the MPLScon Conference and Exhibition. He has published numerous articles on topics relating to data networking and the Internet and is a frequent speaker on networking-related topics at many industry conferences. He holds a bachelor's degree in management information systems from Radford University and an MBA from George Mason University. He is also a Certified Information Systems Security Professional (CISSP).

Dedications

I dedicate this book to my wonderful family, who spent many nights and weekends alone to help me finish the manuscript. To my lovely wife, Roula, I promised you after the IRA book that I wouldn't write another book. Sorry I lied. Thank you for supporting me. To my sons, Joe and Jason, I love you both for the sacrifices you had to make during the last year for me to finish this book.

Acknowledgments

I would like to acknowledge many individuals who made this book possible. Many thanks to Giles Heron from PacketExchange for his thorough review of the material and to his many contributions to the Metro Ethernet space. I would like to thank Irwin Lazar, Mike Bernico, Mark Gallo, and Saaed Sardar for their contributions and for keeping me honest. Thanks to Andrew Malis for his initial work on this project. I also would like to thank many of the authors of the IETF RFCs and IETF drafts whose information has been used for some of the concepts and definitions in this book. This includes the following people: Luca Martini, Nasser El-Aawar, Eric Rosen, and Giles Heron for their work on the encapsulation of Ethernet frames over IP/MPLS networks. V. Kompella, Mark Lasserre, Nick Tingle, Sunil Khandekar, Ali Sajassi, Tom Soon, Yetik Serbest, Eric Puetz, Vasile Radaoca, Rob Nath, Andrew Smith, Juha Heinanen, Nick Slabakov, J. Achirica, L. Andersson, Giles Heron, S. Khandekar, P. Lin, P. Menezes, A. Moranganti, H. Ould-Brahim, and S. Yeong-il for their work on the VPLS draft specification. K. Kompella for his original work on the DTLS draft specification. Special thanks to Daniel O. Awduche for his many contributions to traffic engineering requirements and his phenomenal work in driving multiprotocol lambda switching and GMPLS. Thanks to J. Malcolm, J. Agogbua, M. O'Dell, and J. McManus for their contributions to TE requirements. Many thanks to the CCAMP group and its many contributors to GMPLS, including Peter Ashwood Smith, Eric Mannie, Thomas D. Nadeau, Ayan Banerjee, Lyndon Ong, Debashis Basak, Dimitri Papadimitriou, Lou Berger, Dimitrios Pendarakis, Greg Bernstein, Bala Rajagopalan, Sudheer Dharanikota, Yakov Rekhter, John Drake, Debanjan Saha, Yanhe Fan, Hal Sandick, Don Fedyk, Vishal Sharma, Gert Grammel, George Swallow, Dan Guo, Kireeti Kompella, Jennifer Yates, Alan Kullberg, George R. Young, Jonathan P. Lang, John Yu, Fong Liaw, and Alex Zinin. I would also like to thank the Metro Ethernet Forum and the MPLS Forum for many of their informative references about MPLS and VPLS. I am sure I have missed many of the names of talented people who contributed indirectly to the concepts in this book, many thanks for your efforts.

Last but not least, many thanks to Cisco Systems and the Cisco Press team, John Kane, Dayna Isley, and others for supporting this project.

Contents at a Glance

Table of Contents

Icons Used in This Book

Throughout this book, you see the following icons:

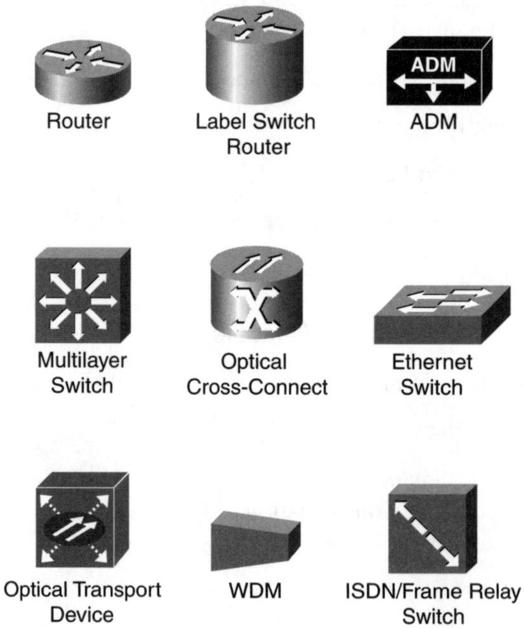

Introduction

Metro Ethernet—opposites attract. Ethernet is a technology that has had major success in the LAN, displacing other once-promising technologies such as Token Ring, FDDI, and ATM. Ethernet's simplicity and price/performance advantages have made it the ultimate winner, extending from the enterprise workgroup closet all the way to the enterprise backbone and data centers. The metro is the last portion of the network standing between subscribers or businesses and the vast amount of information that is available on the Internet. The metro is entrenched with legacy time-division multiplexing (TDM) and SONET/SDH technology that is designed for traditional voice and leased-line services. These legacy technologies are inadequate for handling the bandwidth demands of emerging data applications.

Ethernet in the metro can be deployed as an access interface to replace traditional T1/E1 TDM interfaces. Many data services are being deployed in the metro, including point-to-point Ethernet Line Services and multipoint-to-multipoint Ethernet LAN services or Virtual Private LAN services (VPLS) that extend the enterprise campus across geographically dispersed backbones. Ethernet can run over many metro transport technologies, including SONET/SDH, next-generation SONET/SDH, Resilient Packet Ring (RPR), and wavelength-division multiplexing (WDM), as well as over pure Ethernet transport.

Ethernet, however, was not designed for metro applications and lacks the scalability and reliability required for mass deployments. Deploying Ethernet in the metro requires the scalability and robustness features that exist only in IP and Multiprotocol Label Switching (MPLS) control planes. As such, hybrid Layer 2 (L2) and Layer 3 (L3) IP and MPLS networks have emerged as a solution that marries Ethernet's simplicity and cost effectiveness with the scale of IP and MPLS networks. With many transport technologies deployed in the metro, Ethernet services have to be provisioned and monitored over a mix of data switches and optical switches. It becomes essential to find a control plane that can span both data and optical networks. MPLS has been extended to do this task via the use of the Generalized MPLS (GMPLS) control plane, which controls both data and optical switches. Understanding these topics and more will help you master the metro space and its many intricacies.

Goals and Methods

The goal of this book is to make you familiar with the topic of metro Ethernet—what it is, how it started, and how it has evolved. One thing is for certain: after you read this book, you will never be intimidated by the metro Ethernet topic again. You will be familiar with the different technologies, such as Ethernet switching, RPR, next-generation SONET/SDH, MPLS, and so on, in the context of metro deployments.

The industry today is divided among different pools of expertise—LAN switching, IP routing, and transport. These are three different worlds that require their own special knowledge base. LAN switching expertise is specific to individuals who come from the enterprise space, IP routing expertise is more specific to individuals who deal with public and private IP routed backbones, and transport expertise is specific to individuals who deal with TDM and optical networks. The metro blends all these areas of expertise. This book attempts to bridge the gap between enterprise LAN, IP/MPLS, and transport knowledge in the same way metro bridges the gap between enterprise networks and IP routed backbones over a blend of transport technologies.

The style of this book is narrative. It goes from simple to more challenging within each chapter and across chapters. The big picture is always presented first to give you a better view of what is being described in the chapter, and then the text goes into more details. It is possible to skip the more detailed sections of the book and still have a complete picture of the topic. I call the different levels within a chapter or across chapters "warps." Different readers will find comfort in different warps. The main thing is to learn something new and challenging every time you enter a new warp.

Who Should Read This Book?

The book is targeted at a wide audience, ranging from nontechnical, business-oriented individuals to very technical individuals. The different people who have interest in the subject include network operators, engineers, consultants, managers, CEOs, and venture capitalists. Enterprise directors of technology and CIOs will read the book to assess how they can build scalable virtual enterprise networks. Telecom operators will find in the book a way to move into selling next-generation data services. Engineers will augment their knowledge base in the areas of Ethernet switching, IP/MPLS, and optical networks. Salespeople will gain expertise in selling in a fast-growing metro Ethernet market. Last but not least, businesspeople will understand the topic to the level where they can make wise investments in the metro Ethernet space.

How This Book Is Organized

This book is organized into two main parts:

- Part I—Ethernet: From the LAN to the MAN

  This part of the book—Chapters 1 through 4—starts by describing the different drivers that motivated the adoption of metro Ethernet services and how they have evolved in the United States versus internationally. You will see how Ethernet has moved from the LAN into the MAN and how it is complementing existing and emerging metro technologies such as SONET/SDH, next-generation SONET, RPR, and WDM. You will then learn about the different Ethernet services, such as point-to-point Ethernet Line Services and multipoint-to-multipoint Ethernet LAN services as represented by the concept of Virtual Private LAN Service (VPLS). This part of the book explains the challenges of deploying Ethernet networks and how hybrid Ethernet and IP MPLS networks have emerged as a scalable solution for deploying L2 Ethernet VPN services.

- Part II—MPLS: Controlling Traffic over Your Optical Metro

  MPLS is an important technology for scaling metro deployments. Whereas the first part of the book discusses MPLS in the context of building Layer 2 metro Ethernet VPNs, Part II—Chapters 5 through 8—explores the use of MPLS to control the traffic trajectory in the optical metro. The metro is built with data-switching, SONET/SDH, and optical-switching systems. The act of provisioning different systems and controlling traffic across packet and optical systems is difficult and consitutes a major operational expense. GMPLS has extended the use of MPLS as a universal control plane for both packet/cell and optical systems. GMPLS is one of those "warp 7" subjects. Part II first familiarizes you with the subject of traffic engineering and how the RSVP-TE signaling protocol is used to control traffic trajectory and reroute traffic in the case of failure. This makes the transition into the topic of GMPLS go smoother, with many of the basic traffic engineering in packet/cell networks already defined.

Chapters 1 through 8 and the appendix cover the following topics:

- **Chapter 1, "Introduction to Data in the Metro"**—The metro has always been a challenging environment for delivering data services, because it was built to handle the stringent reliability and availability needs of voice communications. The metro is evolving differently in different regions of the world, depending on many factors. For example, metro Ethernet is evolving slowly in the U.S. because of legacy TDM deployments and stiff regulations, but it is evolving quickly in other parts of the world, especially in Asia and Japan, which do not have as many legacy TDM deployments and are not as heavily regulated.
- **Chapter 2, "Metro Technologies"**—Metro Ethernet services do not necessitate an all-Ethernet Layer 2 network; rather, they can be deployed over different technologies such as next-generation SONET/SDH and IP/MPLS networks. This chapter goes into more details about the different technologies used in the metro.
- **Chapter 3, "Metro Ethernet Services"**—Ethernet over SONET, Resilient Packet Ring, and Ethernet transport are all viable methods to deploy a metro Ethernet service. However, functionality needs to be offered on top of metro equipment to deliver revenue-generating services such as Internet connectivity or VPN services. Chapter 3 starts by discussing the basics of Layer 2 Ethernet switching to familiarize you with Ethernet switching concepts. You'll then learn about the different metro Ethernet services concepts as introduced by the Metro Ethernet Forum (MEF). Defining the right traffic and performance parameters, class of service, and service frame delivery ensures that buyers and users of the service understand what they are paying for and also helps service providers communicate their capabilities.
- **Chapter 4, "Hybrid L2 and L3 IP/MPLS Networks"**—Chapter 4 focuses first on describing a pure Layer 3 VPN implementation and its applicability to metro Ethernet. This gives you enough information to compare Layer 3 VPNs and Layer 2 VPNs relative to metro Ethernet applications. The chapter then delves into the topic of deploying L2 Ethernet services over a hybrid L2 Ethernet and an L3 IP/MPLS network. Some of the basic scalability issues that are considered include restrictions on the number of customers because of the VLAN-ID limitations, scaling the Layer 2 backbone with spanning tree, service provisioning and monitoring, and carrying VLAN information within the network.
- **Chapter 5, "MPLS Traffic Engineering"**—Previous chapters discussed how metro Ethernet Layer 2 services can be deployed over an MPLS network. Those chapters also covered the concept of pseudowires and LSP tunnels. In Chapter 5, you'll learn about the different parameters used for traffic engineering. Traffic engineering is an important MPLS function that allows the network operator to have more control over how traffic traverses its network. This chapter details the concept of traffic engineering and its use.
- **Chapter 6, "RSVP for Traffic Engineering and Fast Reroute"**—MPLS plays a big role in delivering and scaling services in the metro, so you need to understand how it can be used to achieve traffic engineering and protection via the use of Resource Reservation Protocol traffic engineering (RSVP-TE). In this chapter, you see how MPLS, through the use of RSVP-TE, can be used to establish backup paths in the case of failure. This chapter discusses the basics of RSVP-TE and how it can be applied to establish LSPs, bandwidth allocation, and fast-reroute techniques. You'll get a detailed explanation of the RSVP-TE messages and objects to give you a better understanding of this complex protocol.

- **Chapter 7, "MPLS Controlling Optical Switches"**—The principles upon which MPLS technology is based are generic and applicable to multiple layers of the transport network. As such, MPLS-based control of other network layers, such as the TDM and optical layers, is also possible. Chapter 7 discusses why Generalized MPLS (GMPLS) is needed to dynamically provision optical networks. You'll learn about the benefits and drawbacks of both static centralized and dynamic decentralized provisioning models. Chapter 7 also introduces you to the different signaling models (overlay, peer, and augmented) and to how GMPLS uses labels to cross-connect the circuits for TDM and WDM networks.
- **Chapter 8, "GMPLS Architecture"**—Generalized MPLS (GMPLS) attempts to address some of the challenges that exist in optical networks by building on MPLS and extending its control parameters to handle the scalability and manageability aspects of optical networks. This chapter explains the characteristics of the GMPLS architecture, such as the extensions to routing and signaling and the technology parameters that GMPLS adds to MPLS to be able to control optical networks.
- **Appendix, "SONET/SDH Basic Framing and Concatenation"**—This appendix presents the basics of SONET/SDH framing and how the SONET/SDH technology is being adapted via the use of standard and virtual concatenation to meet the challenging needs of emerging data over SONET/SDH networks in the metro. The emergence of L2 metro services will challenge the legacy SONET/SDH network deployments and will drive the emergence of multiservice provisioning platforms that will efficiently transport Ethernet, Frame Relay, ATM, and other data services over SONET/SDH.

PART I

Ethernet: From the LAN to the MAN

This chapter covers the following topics:

- The Metro Network
- Ethernet in the Metro
- The Early Metro Ethernet Movers
- The U.S. Incumbent Landscape
- The International Landscape
- A Data View of the Metro
- Metro Services
- Ethernet Access and Frame Relay Comparison

CHAPTER 1

Introduction to Data in the Metro

The metro, the first span of the network that connects subscribers and businesses to the WAN, has always been a challenging environment for delivering data services because it has been built to handle the stringent reliability and availability needs of voice communications. The metro is evolving differently in different regions of the world depending on many factors, including the following:

- **Type of service provider**—Metro deployments vary with respect to the type of service providers that are building them. While regional Bell operating companies (RBOCs) are inclined to build traditional SONET/SDH metro networks, greenfield operators have the tendency to build more revolutionary rather than evolutionary networks.
- **Geography**—U.S. deployments differ from deployments in Europe, Asia Pacific, Japan, and so on. For example, while many metro deployments in the U.S. are SONET centric, China and Korea are not tied down to legacy deployments and therefore could adopt an Ethernet network faster.
- **Regulations**—Regulations tie to geography and the type of service providers. Europe, for example, has less regulation than the U.S. as far as defining the boundary between a data network and a Synchronous Digital Hierarchy (SDH) network; hence, the adoption of Ethernet over SDH deployments could move faster in Europe than in the U.S.

The Metro Network

The metro is simply the first span of the network that connects subscribers and businesses to the WAN. The different entities serviced by the metro include residential and business customers, examples of which are large enterprises (LEs), small office/home office (SOHO), small and medium-sized businesses (SMBs), multitenant units (MTUs), and multidwelling units (MDUs) (see Figure 1-1).

The portion of the metro that touches the customer is called the *last mile* to indicate the last span of the carrier's network. In a world where the paying customer is at the center of the universe, the industry also calls this span the *first mile* to acknowledge that the customer comes first. An adequate term would probably be "the final frontier" because the last span of the network is normally the most challenging and the most expensive to build and is the final barrier for accelerating the transformation of the metro into a high-speed data-centric network.

Figure 1-1 *The Metro*

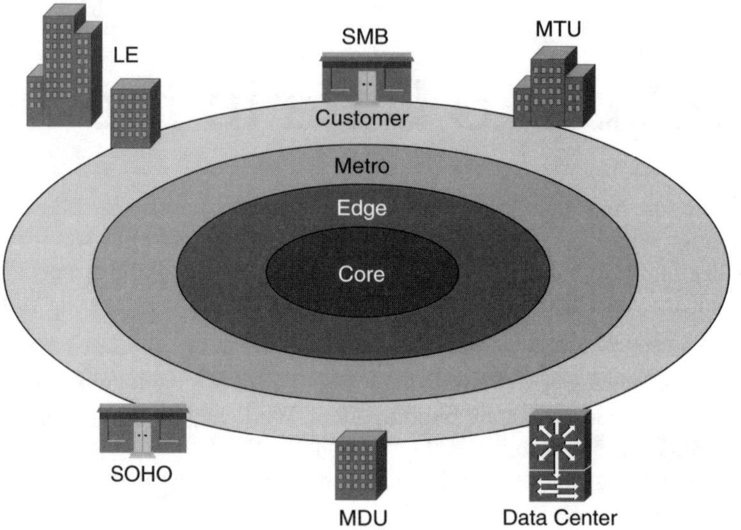

The legacy metro consists primarily of time-division multiplexing (TDM) technology, which is very optimized for delivering voice services. A typical metro network consists of TDM equipment placed in the basement of customer buildings and incumbent local exchange carrier (ILEC) central offices. The TDM equipment consists of digital multiplexers, digital access cross-connects (DACs, often referred to as digital cross-connects), SONET/SDH add/drop multiplexers (ADMs), SONET/SDH cross-connects, and so on.

Figure 1-2 shows a TDM view of a legacy metro deployment. This scenario shows connectivity to business customers for on-net and off-net networks. An *on-net* network is a network in which fiber reaches the building and the carrier installs an ADM in the basement of the building and offers T1 or DS3/OCn circuits to different customers in the building. In this case, digital multiplexers such as M13s multiplex multiple T1s to a DS3 or multiple DS3s to an OCn circuit that is carried over the SONET/SDH fiber ring to the central office (CO). In an *off-net* network, in which fiber does not reach the building, connectivity is done via copper T1 or DS3 circuits that are aggregated in the CO using DACS. The aggregated circuits are cross-connected in the CO to other core COs, where the circuits are terminated or transported across the WAN depending on the service that is being offered.

The operation and installation of a pure TDM network is very tedious and extremely expensive to deploy, because TDM itself is a very rigid technology and does not have the flexibility or the economics to scale with the needs of the customer. The cost of deploying metro networks is the sum of capital expenditure on equipment and operational expenditure. Operational expenditure includes the cost of network planning, installation, operation and management, maintenance and troubleshooting, and so on. What is important to realize is that these operational expenditures could reach about 70 percent of the carrier's total expenditure, which could weigh heavily on the carrier's decision regarding which products and technologies to install in the network.

Figure 1-2 *A TDM View of the Metro*

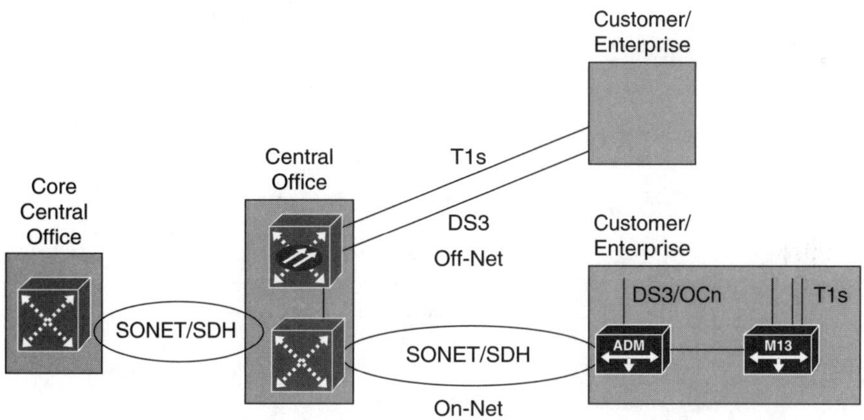

The cost of bringing up service to a customer has a huge effect on the success of delivering that service. The less the carrier has to touch the customer premises and CO equipment to deliver initial and incremental service, the higher the carrier's return on investment will be for that customer. The term *truck rolls* refers to the trucks that are dispatched to the customer premises to activate or modify a particular service. The more truck rolls required for a customer, the more money the carrier is spending on that customer.

The challenge that TDM interfaces have is that the bandwidth they offer does not grow linearly with customer demands but rather grows in step functions. A T1 interface, for example, offers 1.5 Mbps; the next step function is a DS3 interface at 45 Mbps; the next step function is an OC3 interface at 155 Mbps; and so on. So when a customer's bandwidth needs exceed the 1.5-Mbps rate, the carrier is forced to offer the customer multiple T1 (nXT1) circuits or move to a DS3 circuit and give the customer a portion of the DS3. The end effect is that the physical interface sold to the customer has changed, and the cost of the change has a major impact on both the carrier and the customer.

Moving from a T1 interface to an nXT1 or DS3/OCn requires changes to the customer premises equipment (CPE) to support the new interface and also requires changes to the CO equipment to accommodate the new deployed circuits. This will occur every time a customer requests a bandwidth change for the life of the customer connection. Services such as Channelized DS1, Channelized DS3, and Channelized OCn can offer more flexibility in deploying increments of bandwidth. However, these services come at a much higher cost for the physical interface and routers and have limited granularity. This is one of the main drivers for the proliferation of Ethernet in the metro as an access interface. A 10/100/1000 Ethernet interface scales much better from submegabit speeds all the way to gigabit, at a fraction of the cost of a TDM interface.

Figure 1-3 shows the difference between the TDM model and Ethernet model for delivering Internet connectivity. In the TDM model, the metro carrier, such as an ILEC or RBOC, offers the point-to-point T1 circuit, while the ISP manages the delivery of Internet services, which includes managing the customer IP addresses and the router connectivity in the point of presence (POP). This normally has been the preferred model for ILECs who do not want to get involved in the IP addressing and in routing the IP traffic. In some cases, the ILECs can outsource the service or manage the whole IP connection if they want to. However, this model keeps a demarcation line between the delivery of IP services and the delivery of connectivity services.

Figure 1-3 *Connectivity: TDM Versus Ethernet*

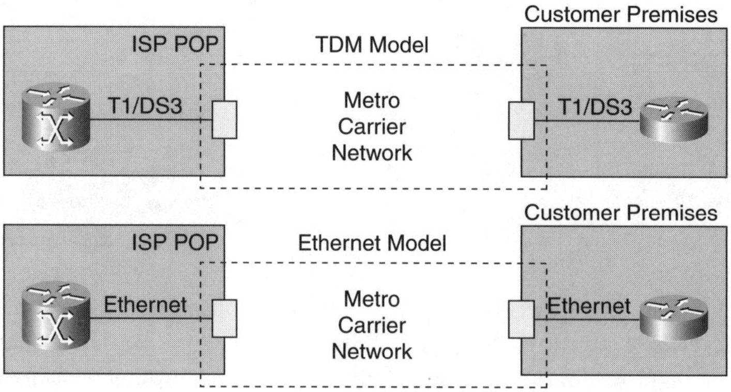

In the Ethernet model, both network interfaces on the customer side and the ISP side are Ethernet interfaces. The ILEC manages the Layer 2 (L2) connection, while the ISP manages the IP services. From an operational perspective, this arrangement keeps the ILEC in a model similar to the T1 private-line service; however, it opens up the opportunity for the ILEC to up-sell additional service on top of the same Ethernet connection without any changes to the CPE and the network.

Ethernet in the Metro

Ethernet technology has so far been widely accepted in enterprise deployments, and millions of Ethernet ports have already been deployed. The simplicity of this technology enables you to scale the Ethernet interface to high bandwidth while remaining cost effective. The cost of a 100-Mbps interface for enterprise workgroup L2 LAN switches will be less than $50 in the next few years.

These costs and performance metrics and Ethernet's ease of use are motivating carrier networks to use Ethernet as an access technology. In this new model, the customer is given an Ethernet interface rather than a TDM interface.

The following is a summary of the value proposition that an Ethernet access line offers relative to TDM private lines:

- **Bandwidth scalability**—The low cost of an Ethernet access interface on both the CPE device and the carrier access equipment favors the installation of a higher-speed Ethernet interface that can last the life of the customer connection. Just compare the cost of having a single installation of a 100-Mbps Ethernet interface versus the installation of a T1 interface for 1.5-Mbps service, a T3 for 45-Mbps service, and an OC3 (155 Mbps) for 100-Mbps service. A TDM interface offering results in many CPE interface changes, many truck rolls deployed to the customer premises, and equipment that only gets more expensive with the speed of the interface.
- **Bandwidth granularity**—An Ethernet interface can be provisioned to deliver tiered bandwidth that scales to the maximum interface speed. By comparison, a rigid TDM hierarchy changes in big step functions. It is important to note that bandwidth granularity is not a function specific to Ethernet but rather is specific to any packet interface. Early deployments of metro Ethernet struggled with this function because many enterprise-class Ethernet switches did not have the capability to police the traffic and enforce SLAs.
- **Fast provisioning**—Deploying an Ethernet service results in a different operational model in which packet leased lines are provisioned instead of TDM circuit leased lines. The packet provisioning model can be done much faster than the legacy TDM model because provisioning can be done without changing network equipment and interfaces. Packet provisioning is a simple function of changing software parameters that would throttle the packets and can increase or decrease bandwidth, establish a connection in minutes, and bill for the new service.

The Early Metro Ethernet Movers

The earliest service providers to move into the metro Ethernet space appeared in the 1999–2000 timeframe in the midst of the telecom bubble and have adopted variations of the same business model across the world.

In the U.S., the early adopters of metro Ethernet were the greenfield service providers that wanted to provide services to some niche segments, such as SMBs that are underserved by the incumbent providers. Other providers have found an opportunity in promoting cheaper bandwidth by selling Ethernet pipes to large enterprises or to other providers such as ISPs or content providers.

The greenfield operators consist of BLECs and metro operators, which are discussed next.

The BLECs

The Building Local Exchange Carriers (BLECs) have adopted a retail bandwidth model that offers services to SMBs which are concentrated in large MTUs. (These are the "tall and shiny buildings" that are usually located in concentrated downtown city areas.) The BLECs focus on wiring the inside of the MTUs for broadband by delivering Ethernet connections to individual offices. The BLECs capitalize on the fact that from the time an SMB places an order, it takes an incumbent operator three to six months to deploy a T1 circuit for that SMB. The BLECs can service the customers in weeks, days, or even hours rather than months and at much less cost.

As shown in Figure 1-4, a BLEC installs its equipment in the basement of the MTU, runs Ethernet in the risers of the building, and installs an Ethernet jack in the customer office. The customer can then get all of its data services from the Ethernet connection.

Figure 1-4 *The BLEC Network Model*

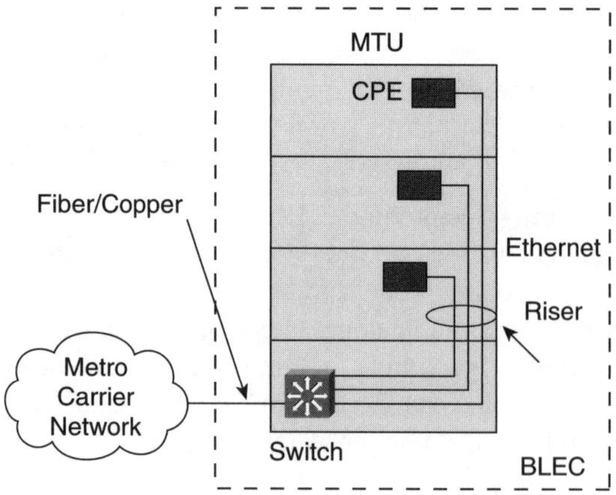

The Metro Ethernet Carrier

Although the BLECs are considered metro operators, they specialize in servicing the MTU customers rather than building connectivity within the metro itself. The metro carriers are focused on building connectivity within the metro and then selling connectivity to BLECs, large enterprises, or even other service providers, depending on the business model. However, a lot of consolidation has occurred because metro operators have acquired BLECs, blurring the distinction between the two different providers.

Whereas some metro carriers have adopted a retail model, selling bandwidth to large enterprises, other metro carriers have adopted a wholesale model, selling bandwidth to other service providers (see Figure 1-5).

Other business plans for metro deployments target cities that want to enhance the quality of life and attract business by tying the whole city with a fiber network that connects schools, universities, businesses, financial districts, and government agencies.

Figure 1-5 *Retail Versus Wholesale Model*

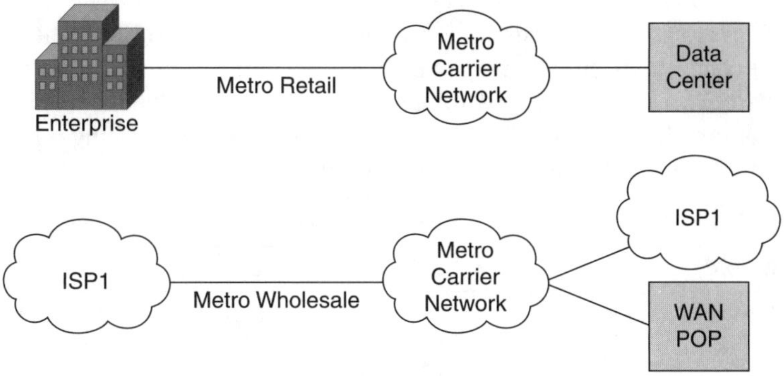

The Greenfield Value Proposition

The following sections describe the value proposition that greenfield operators can offer to attract business away from the incumbents.

Bringing the Service Up in Days Rather Than Months

As mentioned earlier, one of the key selling points for the metro greenfield operators is their ability to bring service up in days. However, to accomplish this, the service has to be almost ready to be brought up once the customer requests it. Greenfields spend a lot of money on idle connections, waiting for a customer to appear.

Pay as You Grow Model

With an Ethernet connection, the customer can purchase an initial amount of bandwidth and SLA and then has the option to change the service in the future by simply calling the provider. The provider could then immediately assign the customer to a different SLA by changing the network parameters via software. Some metro operators offer their customers the ability to change their own bandwidth parameters via a web-based application.

Service Flexibility

With an Ethernet interface, the provider can offer the customer different types of services, such as Internet access, transparent LAN service (TLS), Voice over IP (VoIP), and so on, with minimal operational overhead. Each service is provided over its own virtual LAN (VLAN) and is switched differently in the network. The different services can be sold over the same Ethernet interface or, alternatively, each service can have a separate physical interface.

Lower Pricing Model

The initial claims for the metro Ethernet service were very aggressive. Some of the early marketing campaigns claimed "twice the bandwidth at half the price." The quotes for 100-Mbps Ethernet connections initially ranged from $100 per month to $5000 per month depending on which carrier you talked to and at what time of the day you talked to them. Table 1-1 compares sample pricing for Ethernet and T1/T3 services. The Ethernet pricing might vary widely depending on the region and how aggressive the carrier gets.

Table 1-1 *Sample Pricing Comparison for Ethernet Versus T1/T3 Private-Line Service*

Greenfield	Incumbent
1.5 Mbps at ~$500/month	T1 (1.5 Mbps) at ~$750/month
3 Mbps at ~$750/month	2 * T1 at ~$1500
45 Mbps at ~$2250/month	T3 (45 Mbps) at ~$6000/month

The Challenges of the Greenfield Operators

The BLECs and metro Ethernet carriers have encountered many challenges in their business model that have hindered their success and caused a lot of them to cease to exist after the telecom downturn. This section explores several of those challenges.

The Fight for the Building Riser

Delivering Ethernet connections to the MTU offices requires having access to the building riser, which means dealing with the building owner—although there are regulations that prevent building owners from refusing to allow access to providers. The BLECs, who normally manage to have the first access to the building, have the early field advantage in capturing real estate in the basement and the riser. Of course, how much real estate becomes available or unavailable to other BLECs who are competing for the same MTU usually depends on what percentage of the profits the building owner is receiving.

Cost of Overbuilding the Network

Because many providers in the past operated on the "build it and they will come" theory, millions of dollars were spent on overbuilding the network, which consisted of

- Pulling fiber in the riser
- Building the last-mile connectivity
- Building the core metro network

A challenge for the BLECs is to figure out how much connectivity they need inside the building. Many BLECs have deployed as many connections as possible in the building on the hope that the BLECs will attract customers. This model has, again, resulted in a lot of money spent with no return on investment, forcing many BLECs out of business.

The premise of delivering services to customers in hours and days rather than months is made under the assumption that the BLEC has control of the network facilities inside and outside the building. The perfect solution is to have the BLEC lease or own fiber connections into the building. However, only about five percent of buildings in a metro area have access to fiber, while the rest can only be accessed via copper T1 and DS3 lines. Many BLECs are looking for the "low-hanging fruit," buildings that are already connected via fiber. In many cases, the BLECs try to have arrangements with utility companies to pull fiber into the buildings using existing conduits. In the cases where fiber passes across the building and not into the buildings, the BLECs have to share the cost of digging up the streets with building owners or utility companies. The challenge is that the first BLEC to ask for access into a building has to share the cost of digging the trench, while the BLECs who come after can easily have access to the existing conduit.

For buildings that couldn't have fiber connectivity, the BLECs had to rely on existing copper T1 and DS3 lines to deliver bandwidth into the building. So although the BLECs were competing with the ILECs, they still had to rely on them to provide the copper lines at the ILECs' slow pace of operation.

The metro carriers that are building the metro edge and core infrastructure have sunk a lot of money into buying or leasing the fiber that connects the different points of presence. Many metro providers have locked themselves into multimillion-dollar fiber leases based on the hope that their business will grow to fill up the big pipes.

The Breadth and Reach of Services

Metro carriers have also struggled with the different types of services that they offer and whether the service is offered on a regional or national basis. High-end customers such as large enterprises and financial institutions usually use a one-stop shop: one provider offering local and national connectivity with different types of services, such as Frame Relay or ATM VPN services. An Ethernet-only service approach with no national coverage isn't too attractive. This has forced the metro providers to remain as niche players that do not have the support and reach that the incumbents have.

The Pricing Model

The cheap Ethernet connectivity pricing model could not be sustained. High-speed connections between 10 and 100 Mbps require a higher-speed backbone, which is expensive to build and manage. Also, the greenfield providers were still building up their customer base, and the low Ethernet pricing model did not help with a very small customer base. So Ethernet pricing for 100-Mbps connections was across the map and a trial-and-error process with prices varying by thousands of dollars depending on who you talk to.

The U.S. Incumbent Landscape

While the greenfield operators were fast to build their metro networks, the U.S. incumbents took a sit-and-watch approach to see how the market would shake out. If the greenfield metro

Ethernet model were to succeed, it would start stealing customers from the incumbents, thereby affecting the deployment of their private-line services. Threatened by the newcomers, the RBOCs and IXCs, such as SBC, Verizon, Bellsouth, Qwest, and MCI, initiated requests for information (RFIs) to solicit information from vendors about how to deliver Ethernet services in the metro.

The challenges the incumbents face in deploying metro Ethernet are very different than the challenges of the greenfields. This section discusses some of those challenges, including the following:

- Existing legacy TDM infrastructure
- Building an all-Ethernet data network
- Pricing the services
- Regulations

Existing Legacy TDM Infrastructure

The U.S. metro is entrenched in TDM technology, and billions of dollars have already been spent on building that network. Anyone who intends to build a new service has to consider the existing infrastructure. As inefficient as it may seem, building an Ethernet service over the legacy infrastructure might be the only viable way for some incumbents to make a first entry into the metro Ethernet business. Many of the operational models have already been built for the SONET network. Operators know how to build the network, how to manage and maintain it, and how to deliver a service and bill for it. The incumbents have the challenge of adopting their existing discipline to the metro Ethernet model.

Building an All-Ethernet Data Network

Alternatively, some U.S. incumbents have opted (after many internal debates) to build an all-Ethernet network tailored for data services. However, as of the writing of this book, none of these networks have materialized. Incumbents, who have always dealt with SONET technology, still do not quite understand Ethernet networks. Incumbents normally build their networks and services to tailor to the masses, so any new technology they deploy needs to scale to support thousands of customers nationwide. With Ethernet's roots in enterprise networks, a big gap still exists between what the incumbents need and what existing Ethernet switches, or existing Ethernet standards, have to offer. Incumbents are also unfamiliar with how to manage an Ethernet network, price the service, and bill for it. All of these factors have contributed to the delay in the deployment of such networks.

The deficiencies in Ethernet technology and Ethernet standards in dealing with the metro scalability and availability requirements were one of the main reasons for the proliferation of MPLS in the metro. This topic will be explained in more detail in Chapter 4, "Hybrid L2 and L3 IP/MPLS Networks."

Pricing the Service

For the incumbents, pricing the metro Ethernet services is an extremely challenging exercise. Incumbents that are selling T1 and DS3 connectivity services would be competing with themselves by offering Ethernet services. A very aggressive Ethernet pricing model would jeopardize the sales of T1 and DS3 lines and disrupt the incumbent's business model.

For incumbents selling T1/E1 and DS3 services, their Ethernet pricing model has to do the following to succeed:

- Move hand in hand with existing pricing for legacy services to avoid undercutting the legacy services.
- Offer different levels of services with different price points, in addition to the basic connectivity service. Metro Ethernet services present a good value proposition for both the customer and carrier. The customer can enjoy enhanced data services at higher performance levels, and the carrier can benefit from selling services that it otherwise wouldn't have been able to sell with a simple TDM connection. So the carrier can actually sell the Ethernet connection at a lower price than the legacy connection, based on the hope that the additional services will eventually result in a more profitable service than the legacy services.

The Incumbent Regulations

Another area that challenges the deployment of metro Ethernet services in the U.S. are the regulations that the incumbent carriers have and the delineation between the regulated and unregulated operation inside the same carriers. The regulated portions of the incumbents deal mainly with transport equipment and have rules and guidelines about the use and the location of data switching equipment. The unregulated portion of the incumbent normally has enough flexibility to deploy a mix of hybrid data switching and transport equipment without many ties.

These regulations have created a big barrier inside the incumbents and have created two different operational entities to deal with data and transport. The deployment of new data services such as metro Ethernet will prove to be challenging in the U.S. because such services require a lot of coordination between the data operation and the transport operation of the same incumbent carrier.

The International Landscape

In 2000, while the U.S. market was bubbling with greenfield operators building their metro networks and challenging the almighty RBOCs and IXCs, the metro Ethernet market was taking its own form and shape across the globe. What was different about the rest of the world was the lack of venture capital funding that had allowed new greenfield providers to mushroom in the U.S.

The European Landscape

In Europe, the first activities in metro Ethernet occurred in Scandinavia, specifically Sweden. Telia, the largest Swedish telecom provider, had submitted an RFI for metro Ethernet services. Unlike the U.S., where the providers were focusing on T1 private-line replacement, the target application in Sweden was residential. Many MDU apartment complexes were located in concentrated residential areas, and many of the new developments had fiber already deployed in the basements of the MDUs. Ethernet services seemed like the perfect vehicle to deliver value-added services such as converged voice, data, and video applications. A single Ethernet connection to an MDU could provide Internet access, VoIP, video on demand, and so on.

Also across Europe, a handful of greenfield operators had very aggressive plans to deploy metro Ethernet services, but most faced the same challenges as the U.S. greenfield operators. In pockets of Europe such as Italy, large players such as Telecom Italia were experimenting with an all-Ethernet metro for residential customers.

In general, however, the European metro is entrenched in SDH technology and, like the U.S., has invested in legacy TDM deployments. This puts the big European providers in the same challenging position as the U.S. incumbents in dealing with service cannibalization and the cost of a new buildout. However, Europe differs from the U.S. in that it doesn't have stringent regulations that require a strict boundary between the operation of data switching equipment and SDH transport equipment, which could play a big role in the shift toward metro Ethernet buildouts.

The Asian Landscape

The metro Ethernet landscape in Asia is very different than in the U.S. and Europe. Japan, Korea, and China will prove to be the major players in the deployment of all-Ethernet metro services. One of the major reasons is that these countries haven't invested as much in SONET or SDH and, thus, have a cleaner slate than the U.S. and Europe from which to deploy new data services in the metro.

Many metro Ethernet deployments have already been announced and deployed by big telecom providers such as Korea Telecom SK and others. China will also emerge as a big player in this market after the restructuring of China Telecom into different entities, China Netcom, Unicom, and Railcom.

In Japan, tough competition between telecom providers has driven the cost of private-line services lower than in most other countries. Japan is also a leader in all-metro Ethernet deployments for multimedia services.

A Data View of the Metro

A data view of the metro puts in perspective the different metro services and how they are offered by the different providers.

Figure 1-6 shows a view of the metro with the emphasis on the data access, data aggregation, and service delivery. As you can see, the metro is divided into three segments:

- **Metro access**—This segment constitutes the last-mile portion, which is the part of the network that touches the end customer. For business applications, for example, access equipment resides in a closet in the basement of the enterprise or MTU.
- **Metro edge**—This segment constitutes the first level of metro aggregation. The connections leaving the buildings are aggregated at this CO location into bigger pipes that in turn get transported within the metro or across the WAN.
- **Metro core**—This segment constitutes a second level of aggregation where many edge COs are aggregated into a core CO. In turn, the core COs are connected to one another to form a metro core from which traffic is overhauled across the WAN.

Figure 1-6 *Data View of the Metro*

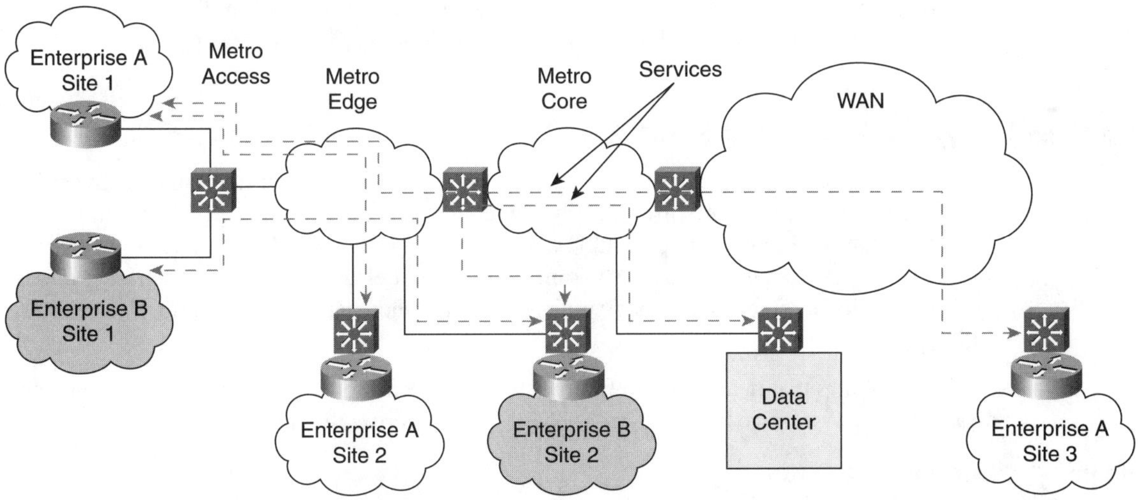

The terminology and many variations of the metro can be confusing. In some cases, there is only one level of aggregation; hence, the building connections are aggregated into one place and then directly connected to a core router. In other scenarios, the metro core CO, sometimes called the metro hub, co-locates with the wide-area POP.

Metro Services

The metro services vary depending on the target market—commercial or residential—and whether it is a retail service or a wholesale service. The following list gives a summary of some of the metro services that are promoted:

- Internet connectivity
- Transparent LAN service (point-to-point LAN to LAN)

- L2VPN (point-to-point or multipoint-to-multipoint LAN to LAN)
- LAN to network resources (remote data center)
- Extranet
- LAN to Frame Relay/ATM VPN
- Storage area networks (SANs)
- Metro transport (backhaul)
- VoIP

Some of these services, such as Internet connectivity and TLS, have been offered for many years. The difference now is that these services are provided with Ethernet connectivity, and the carriers are moving toward a model in which all of these services can be offered on the same infrastructure and can be sold to the same customer without any major operational overhead. This introduces an excellent value proposition to both the customer and the carrier. The services are provisioned through transporting the application over point-to-point or multipoint-to-multipoint L2 connections. The following sections discuss some of these services in greater detail.

LAN to Network Resources

Earlier, in the section "The Metro Network," you saw how Internet service can be delivered by installing at the customer premises an Ethernet connection rather than a T1 TDM connection. After the Ethernet connection is installed at the end customer, the ILEC can sell different services to the customer, such as LAN to network resources. An example of such a service is one that enables an enterprise to back up its data in a remote and secure location for disaster recovery.

Figure 1-7 shows that in addition to Internet service, the customer can have a data backup and disaster recovery service that constantly backs up data across the metro.

Figure 1-7 *LAN to Network Resources*

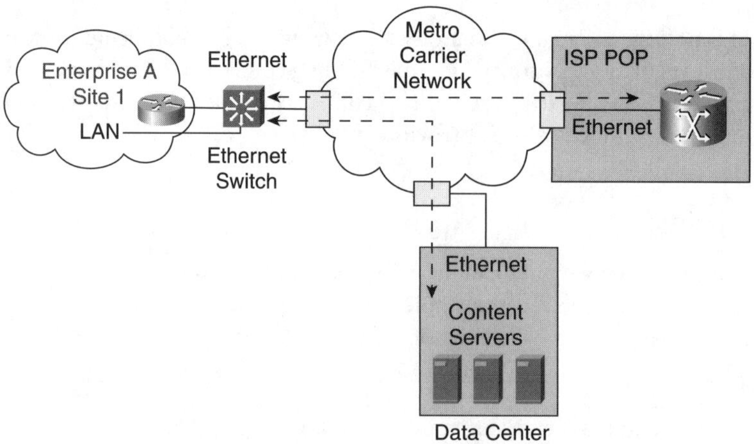

For new data networks in which the connectivity is done via gigabit and 10 gigabit pipes, the metro can be transformed into a high-speed LAN that offers bandwidth-intensive applications that would not normally be feasible to deploy over legacy TDM infrastructure.

As previously mentioned, the service in the metro will take many shapes and forms depending on the target customer. The same LAN to network resources model could be applied toward residential applications, enabling the ILECs to start competing with cable companies in distributing multimedia services. In a residential application, video servers would be located in a metro POP and residential MDU customers could access high-speed digital video on demand over an Ethernet connection. While these services still seem futuristic in the U.S., the international landscape soon could be very different in Europe (particularly Sweden), Japan, and Korea, where the fast deployment of Ethernet networks is already making these applications a reality.

Ethernet L2VPN Services

You may have noticed that many of the services mentioned are pure L2 services that offer connectivity only. This is similar to legacy Frame Relay and ATM services, where the Frame Relay/ATM connection offers a pure L2 pipe and the IP routed services can ride on top of that pipe.

Figure 1-8 shows a carrier deploying an Ethernet L2VPN service. The carrier network behaves as an L2 Ethernet switch that offers multipoint-to-multipoint connectivity between the different customer sites. The customer can benefit from running its own control plane transparently over the carrier's network. The customer routers at the edge of the enterprise could exchange routing protocols without interference with the carrier routing, and the carrier would not have to support any of the customer's IP addressing. An important observation is that while the carrier's network behaves like an L2 Ethernet switch, the underlying technology and the different control planes used in the carrier network are not necessarily based on Ethernet or a Layer 2 control plane.

Figure 1-8 *L2VPN services*

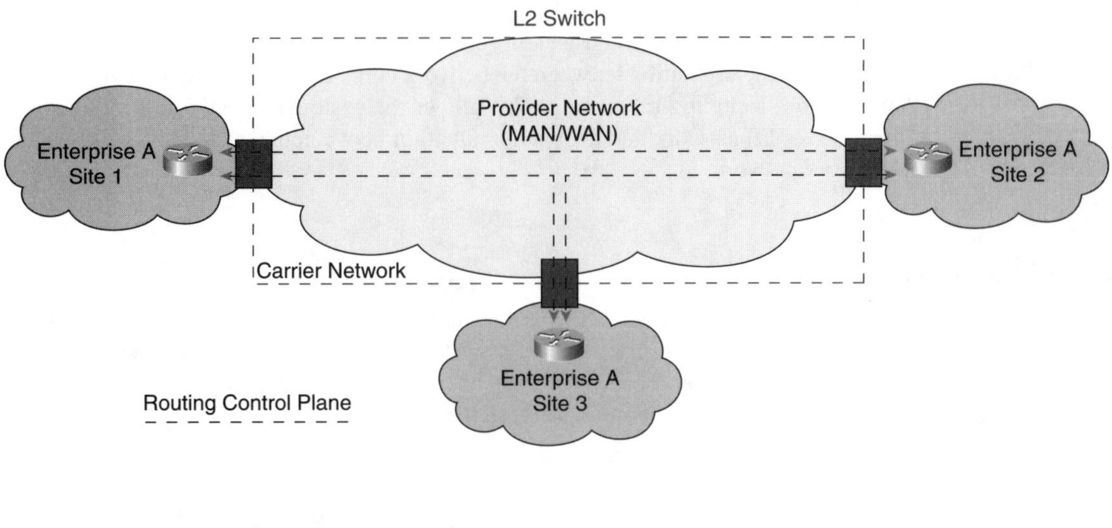

Ethernet Access and Frame Relay Comparison

Frame Relay VPN services have been widely accepted and have proven to be very cost effective compared to point-to-point private-line service. In essence, Ethernet services can be considered the next-generation of Frame Relay because they provide most of the benefits of Frame Relay with better scalability as far as providing higher bandwidth and multipoint-to-multipoint connectivity services. The following list shows some of the similarities and dissimilarities between an Ethernet and a Frame Relay service:

- **Interface speed**—Frame Relay interface speeds range from sub-T1 rates up to OCn speeds. However, Frame Relay has been widely deployed at the lower sub-T1, T1, and DS3 speeds. An Ethernet interface can run at up to 10 Gbps.
- **Last-mile connectivity**—Ethernet services will find better acceptance in on-net deployments (where fiber reaches the building), irrespective of the transport method (as will be explained in the next chapter). Frame Relay has the advantage of being deployed in off-net applications over existing copper T1 and DS3 lines, which so far constitutes a very high percentage of deployments. There are existing efforts in forums, such as the Ethernet in the First Mile (EFM) forum, to run Ethernet directly over existing copper lines. It is unknown at this point whether such a deployment would find acceptance compared to a traditional Frame Relay service.
- **Virtual circuit support**—Both Ethernet and Frame Relay offer a multiplexed interface that allows one customer location to talk to different locations over the same physical interface. The VLAN notion of Ethernet is similar to the Frame Relay permanent virtual circuit (PVC).
- **Multipoint connectivity**—An obvious difference between Frame Relay and Ethernet is that Frame Relay virtual circuits are point-to-point circuits. Any point-to-multipoint or multipoint-to-multipoint connectivity between sites is done via the provisioning of multiple point-to-point PVCs and routing between these PVCs at a higher layer, the IP layer. With Ethernet, the VLAN constitutes a broadcast domain, and many sites can share multipoint-to-multipoint connectivity at L2.
- **L2 interface**—A very important benefit that both Frame Relay and Ethernet offer is the ability to keep the separation between the network connectivity at L2 and the higher-level IP application, including L3 routing. This allows the customer to have control over its existing L2 or L3 network and keep a demarcation between the customer's network and the carrier's network.

Conclusion

The proliferation of data services in the metro is already taking place. You have seen in this chapter how metro data services and specifically Ethernet services are making their move into the metro. The greenfield metro operators have had quite an influence on this shift by putting pressure on traditional metro operators, such as the ILECs. While metro Ethernet is evolving slowly in the U.S. due to legacy TDM deployments and regulations, it has found good success in different parts of the world, especially in Asia and Japan. Metro Ethernet services offer an excellent value proposition both to service providers and to businesses and consumers. Metro Ethernet services will reduce the recurring cost of service deployment while offering much flexibility in offering value-added data services.

Metro Ethernet services do not necessitate an all-Ethernet L2 network; rather, they can be deployed over different technologies such as next-generation SONET/SDH and IP/MPLS networks. Chapter 2, "Metro Technologies," goes into more details about the different technologies used in the metro.

This chapter covers the following topics:

- Ethernet over SONET/SDH (EOS)
- Resilient Packet Ring (RPR)
- Ethernet Transport

CHAPTER 2

Metro Technologies

Metro Ethernet services and applications do not necessarily require Ethernet as the underlying transport technology. The metro can be built on different technologies, such as

- Ethernet over SONET/SDH (EOS)
- Resilient Packet Ring (RPR)
- Ethernet Transport

Ethernet over SONET/SDH

Many incumbent carriers in the U.S. and Europe have already spent billions of dollars building SONET/SDH metro infrastructures. These carriers would like to leverage the existing infrastructure to deliver next-generation Ethernet services. For such deployments, bandwidth management on the network is essential, because of the low capacity of existing SONET/SDH rings and the fact that they can be easily oversubscribed when used for data services.

Incumbents who want to deploy EOS services face tough challenges. Traditionally, for RBOCs and ILECs in the U.S., there is a clear-cut delineation between transport and data. The regulated part of the organization deals with transport-only equipment, not data equipment. With EOS, the equipment vendors blur the line between data and transport, which creates a problem for the adoption of the new technology. So, it is worth spending some time explaining the EOS technology itself.

The benefit of EOS is that it introduces an Ethernet service while preserving all the attributes of the SONET infrastructure, such as SONET fast restoration, link-quality monitoring, and the use of existing SONET OAM&P network management. With EOS, the full Ethernet frame is still preserved and gets encapsulated inside the SONET payload at the network ingress and gets removed at the egress.

As Figure 2-1 shows, the entire Ethernet frame is encapsulated inside an EOS header by the EOS function of the end system at the ingress. The Ethernet frame is then mapped onto the SONET/SDH Synchronous Payload Envelope (SPE) and is transported over the SONET/SDH ring. The Ethernet frame is then extracted at the EOS function on the egress side.

Figure 2-1 *Ethernet over SONET*

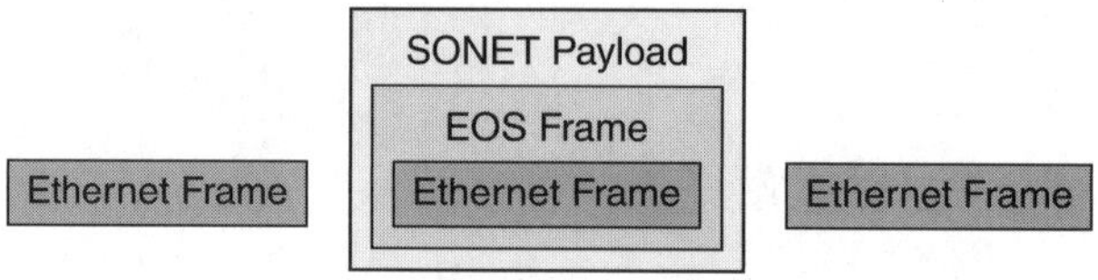

There are two standardized ways to transport Ethernet frames over a SONET/SDH network:

- **LAPS**—Ethernet over the Link Access Procedure SDH is defined by the ITU-T, which published the X.86 standard in February 2001. LAPS is a connectionless protocol similar to High-Level Data Link Control (HDLC).
- **GFP**—Generic Framing Procedure is also an ITU standard that uses the Simple Data Link (SDL) protocol as a starting point. One of the differences between GFP and LAPS is that GFP can accommodate frame formats other than Ethernet, such as PPP, Fiber Channel, fiber connectivity (FICON), and Enterprise Systems Connection (ESCON).

The EOS function can reside inside the SONET/SDH equipment or inside the packet switch. This creates some interesting competing scenarios between switch vendors and transport vendors to offer the Ethernet connection.

Figures 2-2, 2-3, and 2-4 show different scenarios for the EOS connection. In Figure 2-2, the EOS function is inside the ADM. This is normally done via a combination framer/mapper that supports EOS and is placed on a line card or daughter card inside the ADM. The EOS mapping function adds an X.86 or GFP wrapper around the whole Ethernet frame, and the framing function encapsulates the frame in the SONET/SDH SPE. From then on, the SONET/SDH SPE is transported across the SONET/SDH ring and gets peeled off on the egress side. ADMs that contain the EOS function plus other functions such as virtual concatenation (discussed in the next section) are called next-generation ADMs. Figure 2-3 places the EOS function inside the switch.

Figure 2-2 *EOS Function Inside the ADM*

The difference here is that the data equipment and the transport equipment are two different entities that can be owned by different operational groups within the same carrier. This makes it much easier for regulated and unregulated entities within the carrier to deploy a new service. The regulated group's sole responsibility is to provision SONET/SDH circuits, as they would do for traditional voice or leased-line circuits. The unregulated group in turn deploys the higher-layer data services. It is also worth mentioning that in this scenario, the Ethernet switch that delivers the data services has full control of the SONET/SDH tributaries. This is in contrast to Figure 2-2, in which the SONET/SDH tributaries are terminated inside the ADM, and the Ethernet switch sees only a concatenated Ethernet pipe. Figure 2-4 combines the packet-switching, ADM, and EOS functions in the same equipment.

Figure 2-3 *EOS Function Inside the Switch*

For equipment efficiency, this is the optimal solution; however, the deployment of such systems can be challenging if strict operational delineation between packet and transport exists. Such deployments are occurring in the U.S. by smaller competitive telecom providers and by the unregulated portion of the RBOCs/ILECs that do not have many restrictions about data versus transport. Deployments of such systems in Europe are more prevalent because Europe has fewer restrictions than the U.S.

Figure 2-4 *EOS and Switching Functions Inside the ADM*

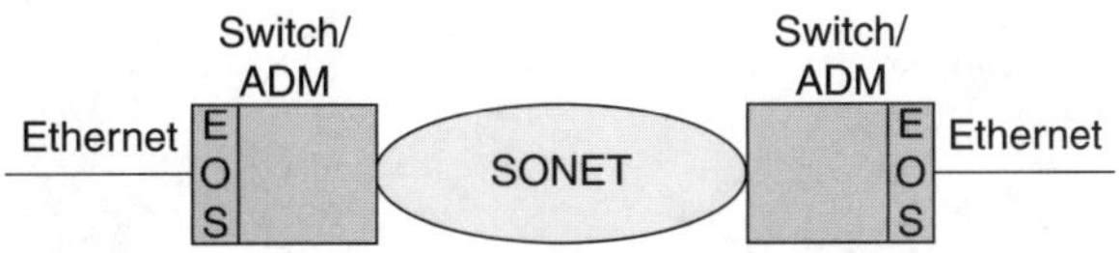

EOS introduces some bandwidth inefficiencies in deploying metro Ethernet services because of the coarse bandwidth granularity of SONET/SDH circuits and the bandwidth mismatch with the sizes of Ethernet pipes. Virtual concatenation (VCAT) is a mechanism used to alleviate such inefficiencies, as discussed next.

The Role of Virtual Concatenation

Virtual concatenation is a measure for reducing the TDM bandwidth inefficiencies on SONET/SDH rings. With standard SONET/SDH concatenation, SONET/SDH pipes are provisioned with coarse granularity that cannot be tailored to the actual bandwidth requirement. The TDM circuits are either too small or too large to accommodate the required bandwidth. On a SONET/SDH ring, once the circuit is allocated, the ring loses that amount of bandwidth whether the bandwidth is used or not.

Appendix A, "SONET/SDH Basic Framing and Concatenation," briefly describes SONET/SDH and the different terminology you see throughout this chapter.

With VCAT, a number of smaller pipes are concatenated and assembled to create a bigger pipe that carries more data per second. Virtual concatenation is done on the SONET/SDH layer (L1) itself, meaning that the different individual circuits are bonded and presented to the upper network layer as one physical pipe. Virtual concatenation allows the grouping of *n* * STS/STM or *n* * VT/VC, allowing the creation of pipes that can be sized to the bandwidth that is needed.

Figure 2-5 highlights the bandwidth efficiency that VCAT can provide. If standard concatenation is used and the bandwidth requirement is for 300 Mbps (about six STS-1s), the carrier has the option of provisioning multiple DS3 interfaces and using packet multiplexing techniques at the customer premises equipment (CPE) to distribute the traffic over the interfaces. (Note that a DS3 interface is the physical interface that runs at a 45-Mbps rate, while an STS-1 is a SONET envelope that can carry 50 Mbps.) Provisioning multiple DS3s at the CPE is normally inefficient, because it increases the cost, does not guarantee the full bandwidth (because of packet load-sharing techniques), and restricts the packet flow to 45 Mbps (because the individual physical circuits are restricted to DS3 bandwidth). The other alternative is for the carrier to allocate a full OC12 (12 STS-1s); this causes the carrier to lose revenue from selling six STS-1s, because they are allocated to a particular customer and cannot be used for other customers on the ring. With virtual concatenation, the carrier can provision a 300 Mbps pipe by bonding six STS-1s as one big pipe—hence no wasted bandwidth.

Figure 2-5 *Virtual Concatenation*

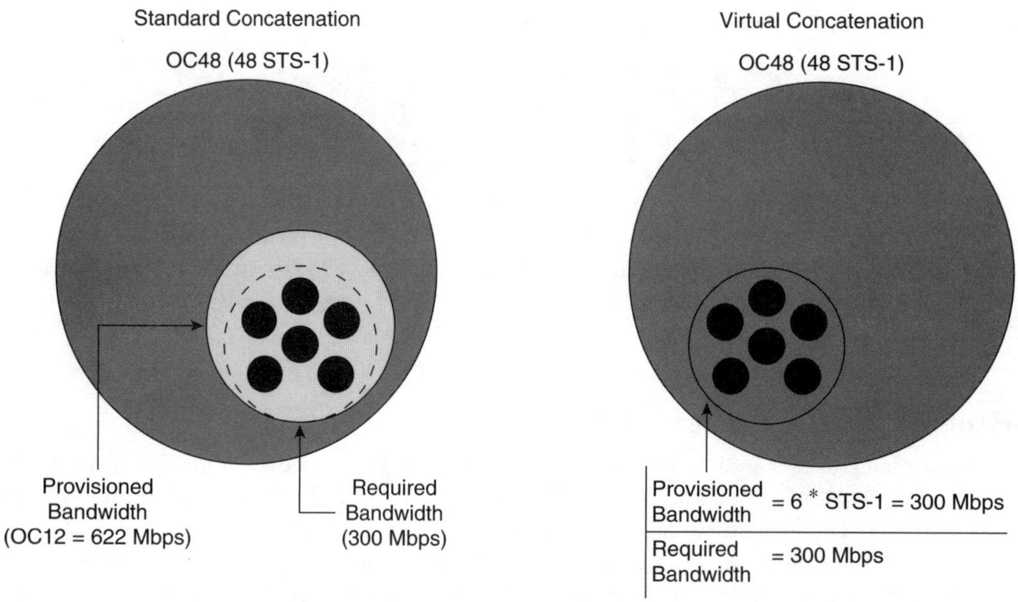

Figure 2-6 shows an example of how multiple services such as Ethernet connectivity services and traditional TDM services can be carried over the same SONET/SDH infrastructure. If the SONET/SDH equipment supports VCAT, a Gigabit Ethernet interface can be carried over a concatenated 21 STS-1 pipe, another Fast Ethernet (FE) 100-Mbps interface can be carried over two STS-1s, and a traditional DS3 interface can be carried over a single STS-1. In many cases, the speed of the Ethernet interface does not have to match the speed on the SONET/SDH side.

A Fast Ethernet 100-Mbps interface, for example, can be carried over an STS-1 (50 Mbps), two STS-1s, or three STS-1s. To handle this oversubscription, throttling of data and queuing of packets or some kind of data backoff need to happen to minimize packet loss.

Figure 2-6 *Transporting Ethernet over SONET*

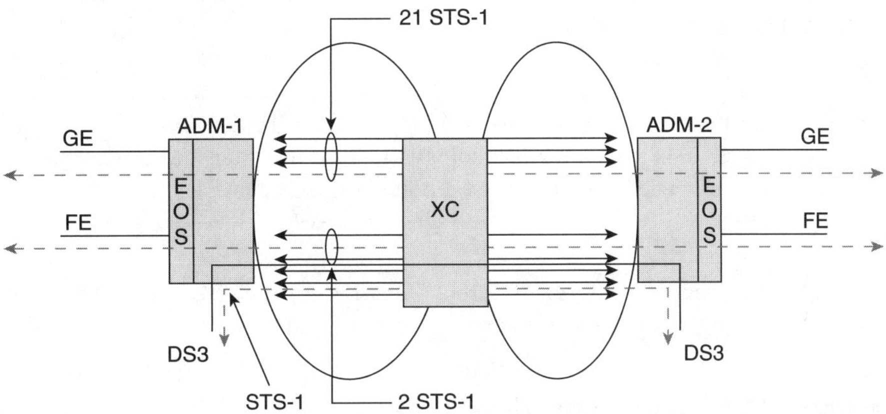

Most rings today support channelization down to the STS-1 (DS3) level and can cross-connect circuits at that level. For T1 services, M13 multiplexers are used to aggregate multiple T1 lines to a DS3 before transporting them on the ring. SONET/SDH equipment that operates at the VT/VC level is starting to be deployed by some RBOCs, which means that with virtual concatenation, circuits of *n* * VT/VC size can be provisioned.

The EOS and VCAT functions are implemented at the entry and exit points of the SONET/SDH infrastructure, and not necessarily at every SONET/SDH station along the way. In Figure 2-6, ADMs 1 and 2 support the EOS and VCAT functions, while the cross-connect (XC) that connects the two rings functions as a traditional cross-connect. However, for VCAT to be effective, the SONET/SDH equipment on the ring has to be able to cross-connect the tributaries supported by the VCAT; otherwise, the bandwidth savings on the ring are not realized. So, if the equipment on the ring supports the allocation of STS-1 circuits and higher, the smallest circuit that can be allocated is an STS-1 circuit. If the terminating equipment supports VCAT to the VT 1.5 level (T1), a full STS-1 bandwidth is still wasted on the ring even if the CPE is allocated *n* * VT 1.5 via VCAT. In Figure 2-6, for example, if ADMs 1 and 2 support VCAT down to the VT 1.5 (T1) level, and the cross-connect can cross-connect only at the STS-1 (DS3) level, the savings are not realized.

Link Capacity Adjustment Scheme

Virtual concatenation is a powerful tool for efficiently grouping the bandwidth and creating pipes that match the required bandwidth. However, the customer bandwidth requirement could change over time, which requires the SONET/SDH pipes to be resized. This could cause network disruption as more SONET/SDH channels are added or removed. Link Capacity Adjustment Scheme (LCAS) is a protocol that allows the channels to be resized at any time without disrupting the traffic or the link. LCAS also performs connectivity checks to allow failed links to be removed and new links to be added dynamically without network disruption.

The combination of EOS, VCAT, and LCAS provides maximum efficiency when deploying Ethernet services over SONET.

NOTE Virtual concatenation and EOS are orthogonal technologies, meaning that they are totally independent. EOS is a mapping technology that runs over standard concatenation and VCAT; however, the full benefits are achieved if done over the latter.

The following sections describe different scenarios in which EOS is used as a pure transport service or is applied in conjunction with packet switching.

EOS Used as a Transport Service

Ethernet over SONET/SDH by itself is still a transport service with an Ethernet interface, similar to the traditional private-line service with a T1, DS3, or OCn interface. EOS offers what is comparable to a point-to-point packet leased-line service. It provides an easy migration for carriers that sell transport to get their feet wet with Ethernet services. EOS is a "packet mapping" technology, not a "packet switching" technology, and does not offer the packet multiplexing that is needed for the aggregation and deaggregation of services. To deliver enhanced switched data services, you need to introduce packet-switching functionality into the metro equipment.

The lack of packet multiplexing for the EOS service and the fact that thousands of point-to-point circuits need to be provisioned between the customers and the central office (CO) create a problem in the aggregation of services in large-scale deployments. Each individual EOS circuit could be presented as a separate Ethernet interface in the CO. With thousands of customers getting an EOS circuit, the CO would have to terminate thousands of individual Ethernet wires. Imagine if the Public Switched Telephone Network (PSTN) were still operating with each customer line terminated inside the CO as a physical wire rather than as a logical circuit. This would create a big patch-panel effect and a nightmare for provisioning and switching between circuits. This patch-panel effect presents a scalability limitation for large-scale EOS deployments, as explained next.

Figure 2-7 shows a scenario in which a carrier is using EOS to sell a basic Internet-connectivity service. A SONET/SDH metro access ring connects multiple enterprise and multitenant unit (MTU) buildings to a CO location. Next-generation ADMs in the basements of the buildings provide 100-Mbps Ethernet connections that connect to the individual routers at each customer premise. The ring itself, in this example, allows channelization down to the VT 1.5 (T1) level, and each Ethernet connection is mapped to one or n * VT 1.5 circuits.

Figure 2-7 *EOS Inside Transport Equipment*

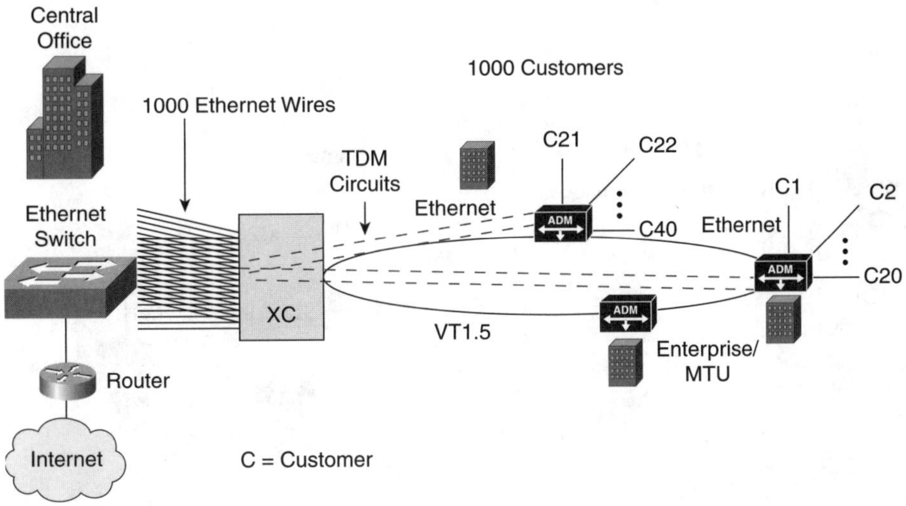

For every customer who is provisioned with an Ethernet interface, an Ethernet interface is extended out of the XC at the CO, because the XC works at the TDM level, and each circuit has to be terminated individually. The individual Ethernet interfaces are then connected to an Ethernet switch that aggregates the traffic toward the ISP router. This means that if a building has 20 customers, 20 different circuits have to be provisioned for that building and have to be terminated in the CO. If the CO supports 50 buildings with 20 customers per building, 1000 TDM circuits have to be provisioned, and hence 1000 Ethernet interfaces have to be terminated in the CO. This model is very inefficient and does not scale well in terms of equipment or management. The XC will be loaded with physical Fast Ethernet interfaces, and the physical connectivity is unmanageable. The logical solution for this problem is to introduce aggregation techniques inside the cross-connect using Ethernet VLANs and to aggregate multiple Ethernet circuits over a single Gigabit or 10 Gigabit Ethernet interface where each circuit is individually identified. While such techniques are possible, they would mean more involvement of transport vendors on the data side, which is challenging from an operational perspective, especially in the U.S.

Figure 2-8 shows an example in which the XC aggregates the different EOS circuits over a single Ethernet interface that connects to an Ethernet switch. For this to happen, the XC needs to be able to logically separate the individual EOS circuits when presenting them to the Ethernet switch. This needs to be done because the traffic sent from the Ethernet switch to the XC over the GE port needs to be tagged with the right circuit ID to reach the correct destination. One method is to have the XC tag individual circuits with a VLAN ID before sending the traffic to the Ethernet switch. Other current implementations put the whole Ethernet bridging function inside the XC itself to allow the multiple EOS streams to be aggregated over a single interface when leaving the transport equipment. This has all the signs of fueling an ongoing industry debate over what functions reside in which equipment as the data vendors and transport vendors start stepping on each other's toes.

Figure 2-8 *EOS Aggregation Inside Transport Equipment*

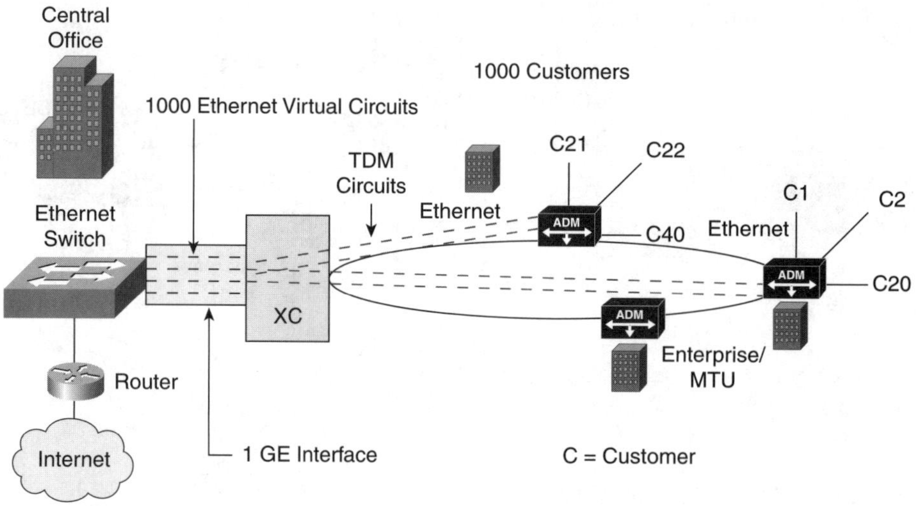

An obvious benefit of a transport service that gives each customer its own TDM circuit is that the customer is guaranteed the full bandwidth. The metro carriers that sell SONET/SDH circuits have dealt with this model for years and know full well how to substantiate the SLAs they sell. When this model is used with VCAT, which enables carriers to tailor the size of the circuit to the customer's need, carriers can realize great bandwidth efficiency and offer firm QoS guarantees. However, you need to weigh this with the complexity of managing the multitude of additional TDM circuits that have to be provisioned, because all these new circuits need to be cross-connected in the network.

EOS with Packet Multiplexing at the Access

The previous example assumes that each customer in the building gets an individual TDM circuit. Another alternative is for the service provider to introduce packet multiplexing in the access switch. The service provider can achieve cost savings by having multiple customers share the same TDM circuit. These cost savings translate into lower cost for the basic connectivity service provided to the customer.

Figure 2-9 shows a scenario where, in the same 50-building metro, each building has an STS-1 (DS3) link that is shared by all 20 customers in each building. This greatly reduces the number of TDM circuits that have to be provisioned, because all customers in the same building share the same STS-1 circuit toward the CO. This reduces the total TDM circuits from 1000 to 50. Notice that 50 Ethernet ports still need to be terminated in the CO if the cross-connect does not have tagging or packet-multiplexing capabilities.

Figure 2-9 *EOS with Packet Multiplexing at the Access*

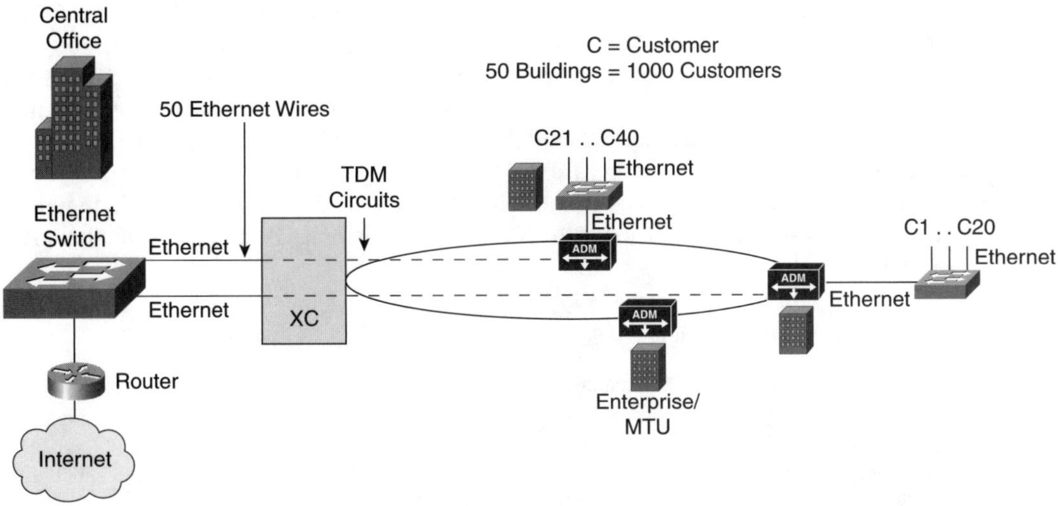

Packet multiplexing in the last mile is yet another incremental step that the metro carriers would have to adopt to move in the direction of delivering switched Ethernet services. Although this model reduces the number of TDM circuits that need to be provisioned, it introduces issues of circuit oversubscription and SLA guarantees. Traffic from multiple customers would be fighting for the STS-1 link, and one customer with a Fast Ethernet (100 Mbps) interface could easily oversubscribe the 45-Mbps link. The carrier would need to use techniques such as traffic policing and traffic shaping to be able to sell its customers tiered bandwidth. These techniques are discussed in Chapter 3, "Metro Ethernet Services," as part of a discussion about bandwidth parameters defined by the Metro Ethernet Forum (MEF).

EOS with Packet Switching

The discussion thus far has addressed the ability to deliver a basic point-to-point leased-line or Internet-connectivity service. More-advanced VPN services can also be delivered over a SONET/SDH metro network that supports EOS. With a VPN service, the assumption is that different locations of the "same" customer exist in a metro area, and the customer wants to be able to tie to these locations via a virtual network. This type of service requires packet switching. Of course, if all the customer wants is a point-to-point service, no switching is required.

Packet switching can be delivered using either of two methods:

- Centralized switching
- Local switching

EOS with Centralized Switching

With centralized switching, a TDM circuit is provisioned from each building to the CO. All circuits are terminated in the CO, which is where the packet switching happens. Note that the standard SONET/SDH operation in unidirectional path switched rings (UPSRs) is to have active circuits and protect circuits on the other side of the ring to achieve the 50-ms ring failure. So, in the metro that has 50 buildings, 50 active STS-1s and 50 protect STS-1 circuits are provisioned. Also note that in case the XC does not support packet tagging or switching, 50 Ethernet interfaces need to be connected to the Ethernet switch at the CO. In Figure 2-10, customer A in sites 1 and 2 belongs to VPN A, while customer B in sites 1 and 2 belongs to VPN B.

Figure 2-10 *EOS with Centralized Switching*

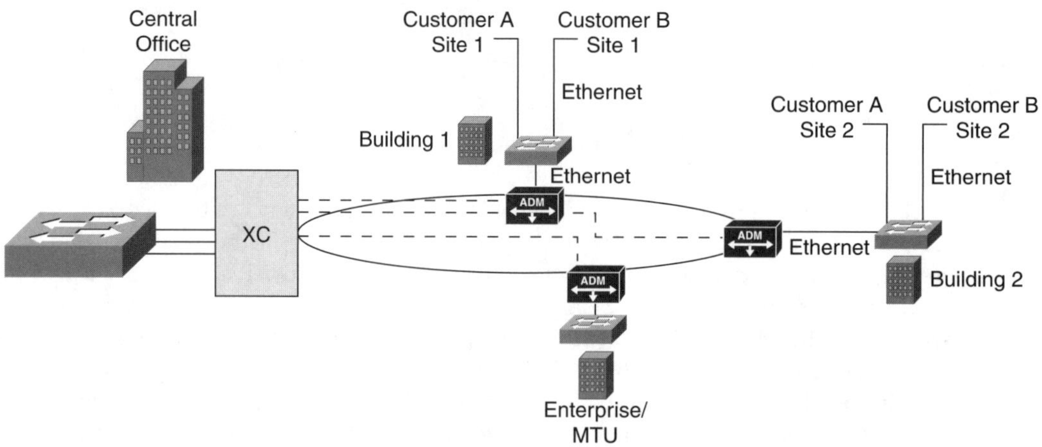

EOS with Local Switching

With local switching, packet switching occurs on each node in the ring. The difference here is that TDM circuits are no longer provisioned between the buildings and the CO but are instead provisioned around the ring. Each ADM in the building terminates circuits for both east and west, and packets get switched at the local switching function in the basement of the building, as shown in Figure 2-11. In this case, SONET/SDH ring protection is not used. The metro carrier must rely on higher-level protection. In the case of L2 Ethernet, this means implementing standard spanning-tree mechanisms, such as the Spanning Tree Protocol (STP), or some other proprietary mechanisms that the Ethernet switch vendor offers. For example, STP would block one side of the ring to prevent a broadcast storm. Also note in Figure 2-11 that in each ADM, a separate Ethernet interface is dedicated to each TDM circuit that gets terminated, unless the ADM itself has a packet-switching function to aggregate the traffic toward the building. If more bandwidth is needed for the building, VCAT can be used to aggregate more circuits while still making them look like a single pipe.

Figure 2-11 *EOS with Local Switching*

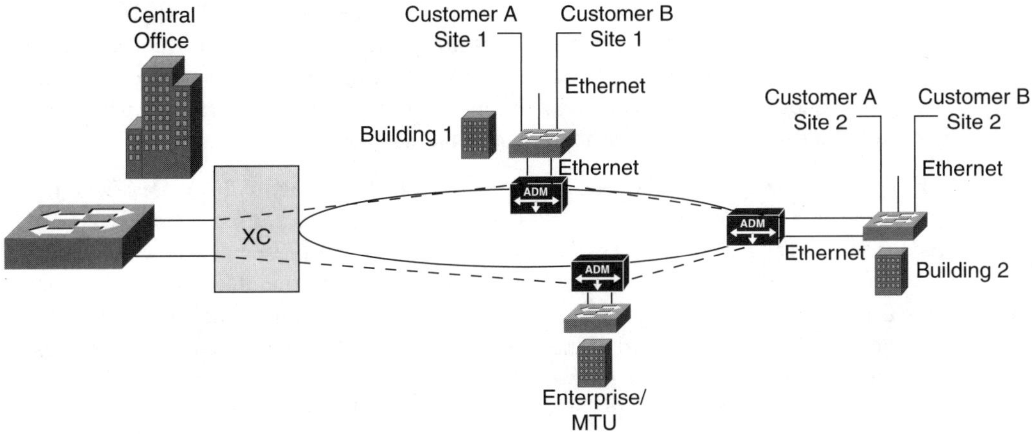

A variation of local switching is to integrate the Ethernet switching function and the ADM/EOS function into one box, as shown in Figure 2-12.

Figure 2-12 *A Variation of EOS with Local Switching*

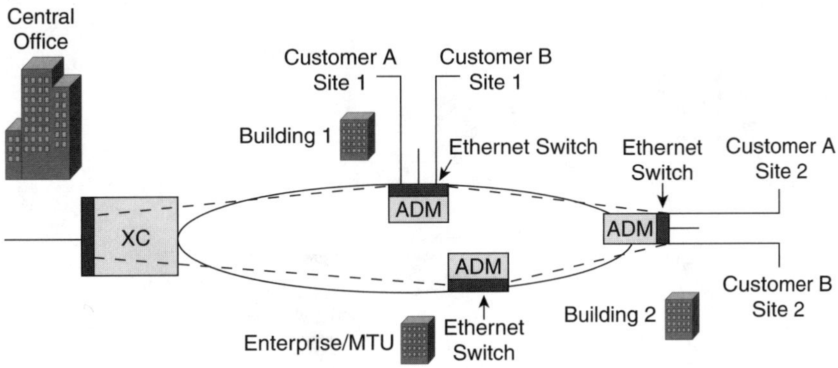

In this case, the TDM circuits are still terminated at each switch/ADM box on the ring. The benefit of this model is that it reduces the number of boxes deployed in the network; however, it blurs the line between the operation of data and TDM networks.

You probably realize by now why metro carriers that are used to SONET/SDH provisioning would like to stay close to the same old circuit-provisioning model. The terms "spanning tree" and "broadcast storms" give metro operators the jitters, because these are enterprise terms that sound very threatening for carriers that are bound to strict SLAs.

EOS Interfaces in the Data Equipment

So far, this chapter has discussed different scenarios for having an EOS interface within the transport equipment. This section discusses the scenario in which the EOS interfaces are part of the data equipment rather than the transport equipment. In this model, the transport equipment does not have to deal with mapping the Ethernet frames carried in the SONET/SDH payload; the data-switching equipment does that instead.

The EOS interfaces inside the data equipment, as shown in Figure 2-13, are SONET/SDH interfaces with a mapping function that maps the EOS frames carried inside the SONET/SDH payload to an Ethernet frame. The Ethernet frame is in turn presented to the switching logic inside the data equipment. As in the case of transport equipment, the EOS interface can support VCAT. The advantage of this model is that the switching function, the EOS function, and the VCAT functions are all in the same box and are decoupled from the TDM box, which may already be installed in the network. This allows the data equipment to have better control over mapping the different data streams over the SONET/SDH circuits. With this model, multipoint-to-multipoint switched Ethernet services can be delivered efficiently while leveraging the existing legacy SONET/SDH infrastructure. This also fits better with the current operational model, in which transport and data are managed separately.

Figure 2-13 *EOS in the Data Equipment*

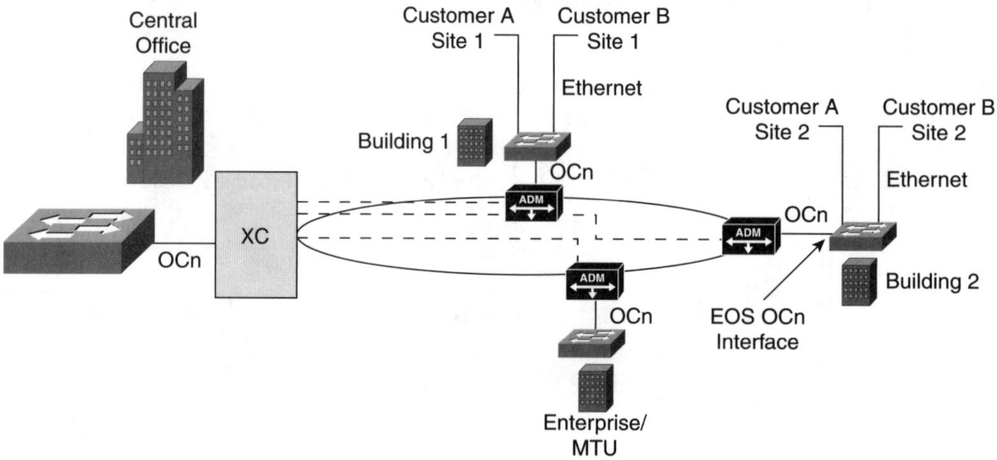

The previously mentioned EOS scenarios are bound to create a lot of debates and confusion in the industry. From a technology perspective, all options are viable, assuming the vendor equipment is capable of delivering the services. From a business perspective, the ownership of the EOS interface determines who makes money on the sale: the data-switching vendors or the transport vendors. You will see numerous debates from both ends about where the EOS services and functions such as VCAT start and terminate.

Resilient Packet Ring

RPR also plays an important role in the development of data services in the metro. RPR is a new Media Access Control (MAC) protocol that is designed to optimize bandwidth management and to facilitate the deployment of data services over a ring network. The roots of RPR go back to the point at which Cisco Systems adopted a proprietary Data Packet Transport (DPT) technology to optimize packet rings for resiliency and bandwidth management. DPT found its way into the IEEE 802.17 workgroup, which led to the creation of an RPR standard that differs from the initial DPT approach.

RPR has so far been a very attractive approach to multiple service operators (MSOs), such as cable operators that are aggregating traffic from cable modem termination systems (CMTSs) in the metro. It remains to been seen whether RPR will be deployed by the incumbent carriers, such as the RBOCs and ILECs, that so far haven't been widely attracted to the RPR concept. The primary reason why they lack interest is that they view RPR deployments as new deployments, compared to EOS deployments, which leverage existing infrastructure and are therefore more evolutionary. RPR is a new packet-ring technology that is deployed over dark fiber or wavelength division multiplexing (WDM) instead of the traditional SONET/SDH rings. RPR could be deployed as an overlay over existing SONET/SDH infrastructure; however, the complexity of overlaying logical RPR rings over physical SONET/SDH rings will probably not be too attractive to many operators. Although RPR and EOS solve different issues in the metro (EOS solves Ethernet service deployment, and RPR solves bandwidth efficiency on packet rings), both technologies will compete for the metro provider's mind share.

Figure 2-14 shows a typical RPR deployment with a cable operator. The CMTSs aggregate the traffic coming over the coaxial cable from businesses and homes and hand over the data portion (assuming the cable is carrying voice/video as well) to the RPR router. Multiple RPR routers connect via an OC48 packet ring, and the traffic gets aggregated in the core hub, where Internet connectivity is established.

Figure 2-14 *RPR Deployments*

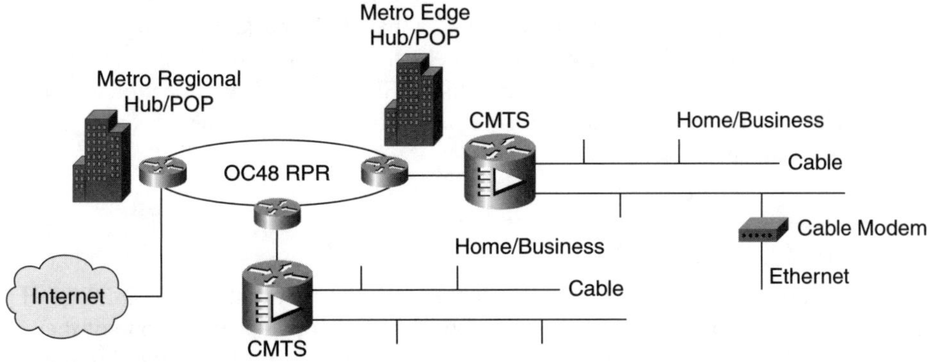

RPR is somehow more commonly associated with routers than with switches, whereas EOS is more commonly associated with switches than routers. The reason for these associations is that DPT historically has been deployed using Cisco IP routers for delivering routed IP services over a packet ring. While the IEEE 802.17 standards body would like to make RPR independent of Layer 2 (L2) switching or Layer 3 (L3) routing, the fact remains that RPR has so far been adopted for L3 services. Also, many routers lack the right functionality to deliver L2 services, which makes EOS more suitable for switches. Again, while the technologically savvy reader might argue that L2 or L3 could be delivered with either technology—and there are existing platforms that support both L2 and L3 services—service provider adoption will be the determining factor in how each technology will most likely be used.

In comparing RPR with traditional SONET/SDH rings, you will realize that RPR deployments have many advantages simply because RPR is a protocol built from the ground up to support data rings. The following sections discuss several features of RPR.

RPR Packet Add, Drop, and Forward

The RPR operation consists of three basic operations: add, drop, and forward. These operations mimic the add/drop mechanisms that are used in traditional SONET networks, where circuits get added, dropped, and cross-connected inside a ring.

The advantage that RPR has over a traditional Ethernet switched packet ring is that Ethernet 802.3 MAC operation processes packets at each node of the ring irrespective of whether the packet destination is behind that node. In contrast, RPR 802.17 MAC forwards the traffic on the ring without doing any intermediary switching or buffering if the traffic does not belong to the node. This reduces the amount of work individual nodes have to do.

In the RPR operation shown in Figure 2-15, traffic that does not belong to a particular node is transited (forwarded) on the ring by the 802.17 MAC. In the Ethernet 802.3 MAC operation, the traffic is processed and buffered at each node for the switching function to determine the exit interface.

RPR's advantage over a SONET/SDH ring is that all the packets coming into the ring share the full-ring bandwidth, and the RPR mechanism manages the bandwidth allocation to avoid congestion and hot spots. In a SONET/SDH ring, TDM time slots are allocated to each circuit, and the bandwidth is removed from the ring whether there is traffic on that circuit or not.

RPR Resiliency

RPR offers ring protection in 50 ms, comparable with the traditional SONET/SDH protection. RPR fast protection with full-ring bandwidth utilization is probably one of the main assets that RPR has when compared to SONET/SDH and other packet-protection mechanisms.

RPR protection is achieved in two ways:

- **Ring wrapping**—The ring is patched at the location of the fault.
- **Ring steering**—In case of failure, the traffic is redirected (steered) at the source toward the working portion of the ring.

Figure 2-15 *RPR Add, Drop, and Forward*

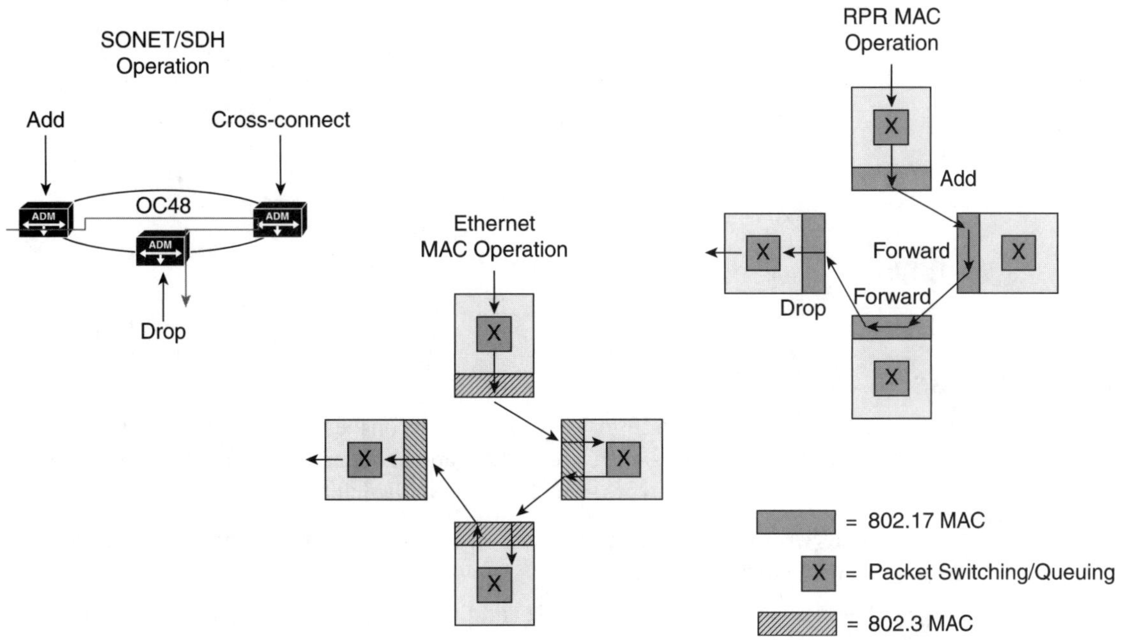

In general, the physical layer detects faults and signals that information to the MAC layer. If the failure is determined to be critical, each affected RPR node initiates a fail-over action for the service flows it originates that are affected by the facility outage. The fail-over action is a simple redirection of the traffic from the failed path to the protection path. The process of alarm notification and redirecting traffic is completed within 50 ms of the outage.

Figure 2-16 compares and contrasts RPR and SONET/SDH. In the SONET/SDH UPSR schemes, for example, 50-ms protection is achieved by having a working fiber and a standby protect fiber at all times. A sending node transmits on both fibers (east and west) at the same time, and a receiving node accepts traffic from only one side. In case of a fiber cut, recovery is done in less than 50 ms. In UPSRs, only 50 percent of the fiber capacity is used, because the other half is kept for failure mode. In RPR, both fiber rings—the outer ring and the inner ring—are used to utilize 100 percent of the ring capacity. In case of a failure, the ring wraps, isolating the failed portion. So, in essence, the effective bandwidth of an RPR ring is twice as much as a SONET/SDH ring because of the SONET/SDH protection.

Figure 2-16 *RPR Protection*

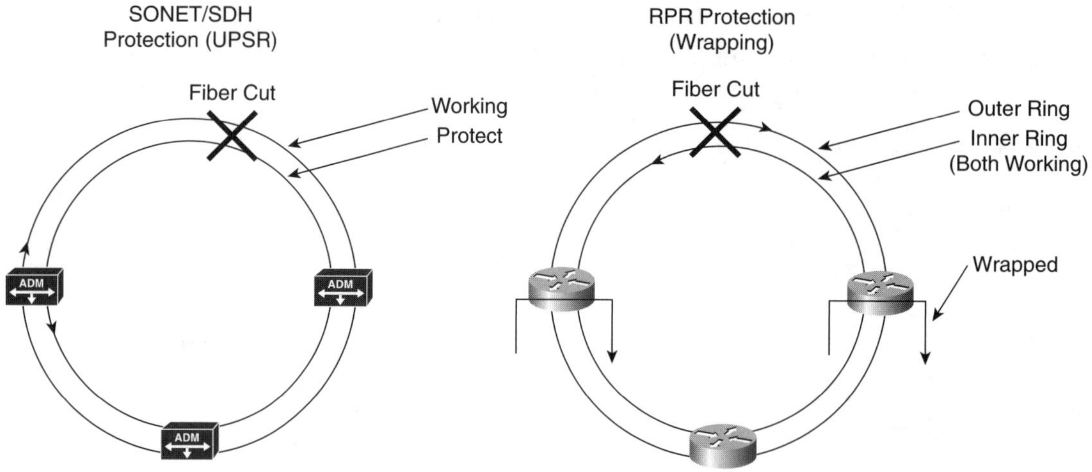

RPR Fairness

RPR implements a fairness algorithm to give every node on the ring a fair share of the ring. RPR uses access-control mechanisms to ensure fairness and to bound latency on the ring. The access control can be broken into two types, which can be applied at the same time:

- **Global access control**—Controls access so that every node can get a fair share of the ring's global bandwidth.
- **Local access control**—Gives the node additional ring access—that is, bandwidth beyond what was globally allocated—to take advantage of segments that are less-used.

RPR uses the special reuse protocol (SRP), which is a concept used in rings to increase the ring's overall aggregate bandwidth. This is possible because multiple spans of the ring can be used simultaneously without having the traffic on one span affect the traffic on the other spans. If a node experiences congestion, it notifies the upstream nodes on the ring, which in turn adjust the transmit rate to relieve downstream congestion.

It helps to contrast ring bandwidth fairness between RPR and L2 Ethernet rings. In the case of an Ethernet ring with L2 switching, there is no such thing as ring fairness, because the QoS decisions are local to each node, irrespective of what is on the ring. You can use rate-limiting techniques to prevent a set of customers who are coming in on one node from oversubscribing the ring; however, it would be hard to have a global fairness mechanism without resorting to complicated QoS management software applications that would coordinate between all nodes.

Figure 2-17 shows three different scenarios for SONET/SDH UPSR, RPR, and L2 Ethernet rings. In the SONET/SDH case, if an STS-1 is allocated, the ring loses an STS-1 worth of bandwidth, irrespective of actual traffic. In the Ethernet case, traffic from A to C and from B to C might oversubscribe the capacity of the point-to-point link between switches SW2 and SW3. In the RPR case, the MAC entity on each node monitors the utilization on its immediate links and makes that information available to all nodes on the ring. Each node can then send more data or throttle back.

Figure 2-17 *Ring Bandwidth*

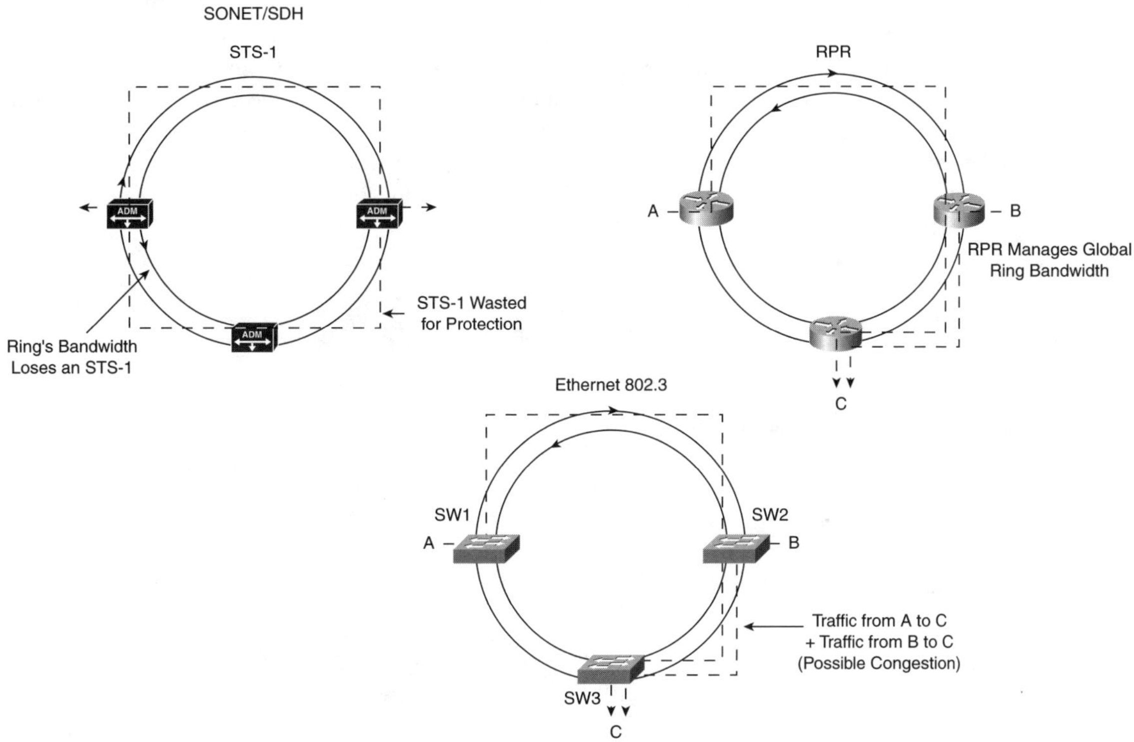

Ethernet Transport

So far, this book has addressed the reasoning behind adopting Ethernet as an access interface rather than a TDM interface. But as discussed in this section, Ethernet isn't limited to being an access technology. Many efforts have been made to extend Ethernet itself into the MAN as a transport technology. Since the early 2000s, metro Ethernet deployments have taken many shapes and forms; some have proven to work, and others have not. When Ethernet is used as a transport technology, the access network can be built in either ring or hub-and-spoke topologies. These are discussed next.

Gigabit Ethernet Hub-and-Spoke Configuration

In a Gigabit Ethernet hub-and-spoke configuration, Ethernet switches deployed in the basement of buildings are dual-homed into the nearest point of presence (POP) or CO. Dedicated fiber, or dedicated wavelengths using WDM, is used for connectivity. Although this is the most expensive approach for metro access deployments because of the cost of fiber, some carriers consider it to be the better solution as far as survivability and scalability compared to deploying Ethernet in a ring topology (described in the next section). With the hub-and-spoke model, the bandwidth dedicated to each building can scale, because the full fiber is dedicated to the building. Protection schemes can be achieved via mechanisms such as link aggregation 802.3ad or dual homing. With link aggregation, two fibers are aggregated into a bigger pipe that connects to the CO. Traffic is load-balanced between the two fibers, and if one fiber is damaged, the other absorbs the full load. This, of course, assumes that the two fibers are run into two different conduits to the CO for better protection. This scenario is shown in Figure 2-18 for the connection between building 1 and the CO.

Figure 2-18 *Ethernet Hub and Spoke*

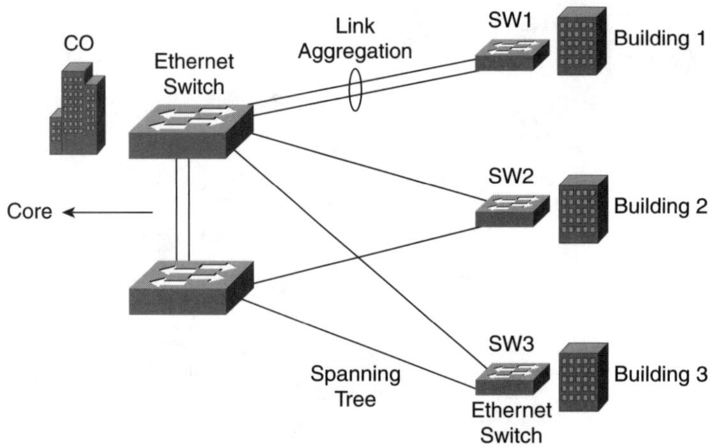

Another approach is to dual-home the fiber into different switches at the CO, as shown in Figure 2-18 for buildings 2 and 3. Although this prevents a single point of failure on the switching side, it creates more complexities, because STP must be run between the buildings and the CO, causing traffic on one of the dual-homed links to be blocked.

Gigabit Ethernet Rings

Many fiber deployments in the metro are laid in ring configurations. Consequently, ring topologies are the most natural to implement and result in cost savings. However, the situation differs depending on whether you are dealing with U.S. carriers or international carriers, incumbents, or greenfields. Ring deployments could be extremely cost-effective for one carrier but a nonissue for another. For existing fiber laid out in a ring topology, Gigabit Ethernet rings are a series of point-to-point connections between the switches in the building basements and the CO, as shown in Figure 2-19. As simple as they might look, Gigabit Ethernet rings may create many issues for the operators because of protection and bandwidth limitations. First of all, ring capacity could be a major issue. Gigabit Ethernet rings have only 1 GB of capacity to share between all buildings, and some of that capacity is not available because spanning tree blocks portions of the ring to prevent loops.

Figure 2-19 *Gigabit Ethernet Rings*

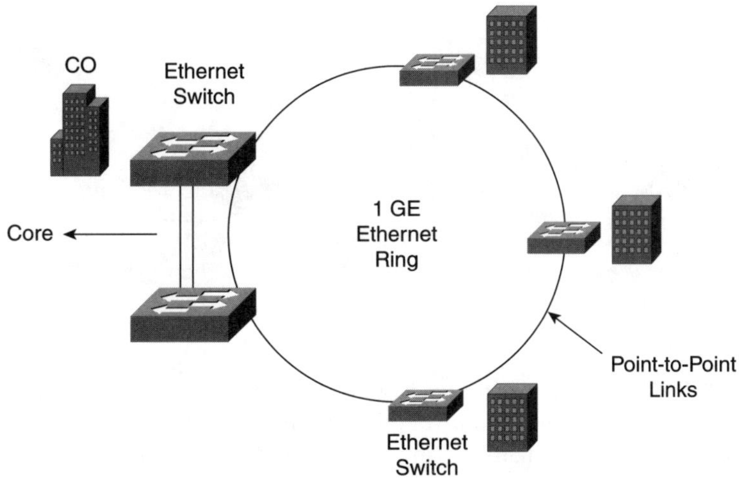

With an Ethernet L2 switched operation, the ring itself becomes a collection of point-to-point links. Even without a fiber cut, spanning tree blocks portions of the ring to prevent broadcast storms caused by loops (see Part A of Figure 2-20). Broadcast storms occur, for example, when a packet with an unknown destination reaches a node. The node floods the packet over the ring according to standard bridging operation as defined in 802.3d. If there is a loop in the network (in this case, the ring), the packet could end up being received and forwarded by the same node over and over. The spanning-tree algorithm uses control packets called bridge protocol data units (BPDUs) to discover loops and block them. Spanning tree normally takes between 30 and 60 seconds to converge. The new 802.1W Rapid Spanning Tree allows faster convergence but still doesn't come close to 50 ms. Many proprietary algorithms have been introduced to achieve ring convergence in less than 1 second, which many operators view as good enough for data services and even for Voice over IP (VoIP) services. However, because L2 switching cannot operate in a loop environment, many of these algorithms still need to block redundant paths in the ring to prevent broadcast storms, and are not considered as reliable as RPR or SONET/SDH protection mechanisms that are more carrier-class. When a fiber cut occurs, spanning tree readjusts, and the new path between the different nodes is established, as shown in Part B of Figure 2-20.

Figure 2-20 *Gigabit Ethernet Rings—Spanning Tree*

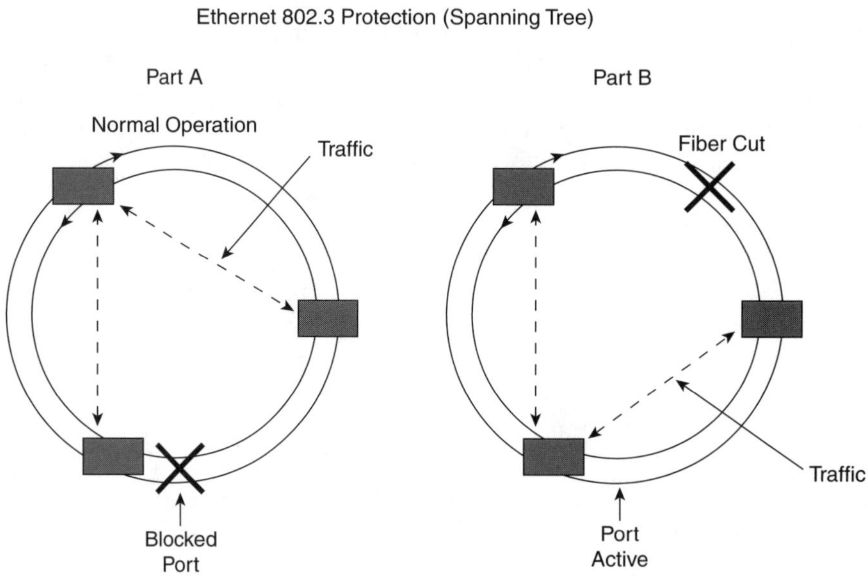

Although 10-Gigabit Ethernet rings would alleviate the congestion issues, initial solutions for 10-GE switches are cost-prohibitive. Initial equipment with 10-GE interfaces was designed for core networks rather than building access. As 10-GE solutions mature and their prices are reduced to fit the building access, 10-GE rings will become a viable solution.

Other methods, such as deploying WDM, could be used to add capacity on the ring. It is debatable whether such methods are cost-effective for prime-time deployments, because they increase the operational overhead of deploying the access ring.

Conclusion

So far, you have read about different technologies that can be used for physical metro connectivity. Ethernet over SONET, RPRs, and Ethernet transport are all viable methods to deploy a metro Ethernet service. Ethernet over SONET presents a viable solution for deploying Ethernet services over an existing installed base. You have seen how virtual concatenation allows better efficiency and bandwidth granularity when mapping Ethernet pipes over SONET/SDH rings. RPR is a packet-ring technology that is attracting much interest from MSOs because it solves many of the restoration and bandwidth inefficiencies that exist in SONET/SDH rings. Ethernet as a transport technology is also a simple and efficient way to deploy Ethernet services; however, by itself, this solution inherits many of the deficiencies of L2 switched Ethernet networks.

Much functionality still needs to be offered on top of metro equipment to deliver revenue-generating services such as Internet connectivity and VPN services. Ethernet, for example, has always been used in a single-customer environment, such as an enterprise network. It is now moving to a multicustomer environment in which the same equipment delivers services to different customers over a shared carrier network. Issues of virtualization of the service and service scalability become major issues. Ethernet over MPLS (EoMPLS) is becoming a viable solution for deploying scalable metro Ethernet services. The MPLS control plane delivers most of the functionality that is lacking in Ethernet switched networks as far as scalability and resiliency. Chapter 3 discusses metro Ethernet services and Layer 2 switching, in preparation for Chapter 4, which discusses delivering Ethernet over hybrid Ethernet and IP/MPLS networks.

This chapter covers the following topics:

- L2 Switching Basics
- Metro Ethernet Services Concepts
- Example of an L2 Metro Ethernet Service
- Challenges with All-Ethernet Metro Networks

CHAPTER 3

Metro Ethernet Services

As discussed in Chapter 1, "Introduction to Data in the Metro," Ethernet services can take either of two forms: a retail service that competes with traditional T1/E1 private-line services, or a wholesale service where a carrier sells a big Ethernet transport pipe to another, smaller service provider. In either case, multiple customers share the same metro infrastructure and equipment. For TDM deployments, sharing the infrastructure is a nonissue, because the services are limited to selling transport pipes, and each customer is allocated a circuit that isolates its traffic from other customers. The customer gets well-defined SLAs, mainly dictated by the circuit that is purchased.

When packet multiplexing and switching are applied, such as in the cases of switched EOS, Ethernet Transport, and RPR, things change. Packets from different customers are multiplexed over the same pipe, and the bandwidth is shared. No physical boundaries separate one customer's traffic from another's, only logical boundaries. Separation of customer traffic and packet queuing techniques have to be used to ensure QoS. Multiple functions have to be well-defined to offer a service:

- How to identify different customers' traffic over a shared pipe or shared network
- How to identify and enforce the service given to a particular customer
- How to allocate certain bandwidth to a specific customer
- How to "transparently" move customers' traffic between different locations, such as in the case of transparent LAN services
- How to scale the number of customers
- How to deploy a VPN service that offers any-to-any connectivity for the same customer

This chapter starts by discussing the basics of L2 Ethernet switching to familiarize you with Ethernet-switching concepts. Then it discusses the different metro Ethernet service concepts as introduced by the Metro Ethernet Forum (MEF).

L2 Switching Basics

L2 switching allows packets to be switched in the network based on their Media Access Control (MAC) address. When a packet arrives at the switch, the switch checks the packet's destination MAC address and, if known, sends the packet to the output port from which it learned the destination MAC.

The two fundamental elements in Ethernet L2 switching are the MAC address and the virtual LAN (VLAN). In the same way that IP routing references stations on the networks via an L3 IP address, Ethernet L2 switching references end stations via the MAC address. However, unlike IP, in which IP addresses are assigned by administrators and can be reused in different private networks, MAC addresses are supposed to be unique, because they are indicative of the hardware itself. Thus, MAC addresses should not be assigned by the network administrator. (Of course, in some cases the MAC addresses can be overwritten or duplicated, but this is not the norm.)

Ethernet is a broadcast medium. Without the concept of VLANs, a broadcast sent by a station on the LAN is sent to all physical segments of the switched LAN. The VLAN concept allows the segmentation of the LAN into logical entities, and traffic is localized within those logical entities. For example, a university campus can be allocated multiple VLANs—one dedicated for faculty, one dedicated for students, and the third dedicated for visitors. Broadcast traffic within each of these VLANs is isolated to that VLAN.

Figure 3-1 shows the concept of an Ethernet LAN using a hub (Part A) and an Ethernet switch (Part B). With an Ethernet hub, all stations on the LAN share the same physical segment. A 10-Mbps hub, for example, allows broadcast and unicast traffic between the stations that share the 10-Mbps bandwidth. The switched LAN on the right allows each segment a 100-Mbps connection (for this example), and it segments the LAN into two logical domains, VLAN 10 and VLAN 20. The concept of VLANs is independent of the stations themselves. The VLAN is an allocation by the switch. In this example, ports 1 and 2 are allocated to VLAN 10, while ports 3 and 4 are allocated to VLAN 20. When stations A1 and A2 send traffic, the switch tags the traffic with the VLAN assigned to the interface and makes the switching decisions based on that VLAN number. The result is that traffic within a VLAN is isolated from traffic within other VLANs.

Ethernet switching includes the following basic concepts:

- MAC learning
- Flooding
- Using broadcast and multicast
- Expanding the network with trunks
- VLAN tagging
- The need for the Spanning Tree Protocol (STP)

MAC Learning

MAC learning allows the Ethernet switch to learn the MAC addresses of the stations in the network to identify on which port to send the traffic. LAN switches normally keep a MAC learning table (or a bridge table) and a VLAN table. The MAC learning table associates the MACs/VLANs with a given port, and the VLAN table associates the port with a VLAN. In Figure 3-1, Part B, the switch has learned the MAC addresses of stations A1, A2, B1, and B2 on ports 1, 2, 4, and 3, respectively. It also shows that ports 1 and 2 are associated with VLAN 10 and ports 3 and 4 are associated with VLAN 20.

Figure 3-1 *Ethernet MACs and VLANs*

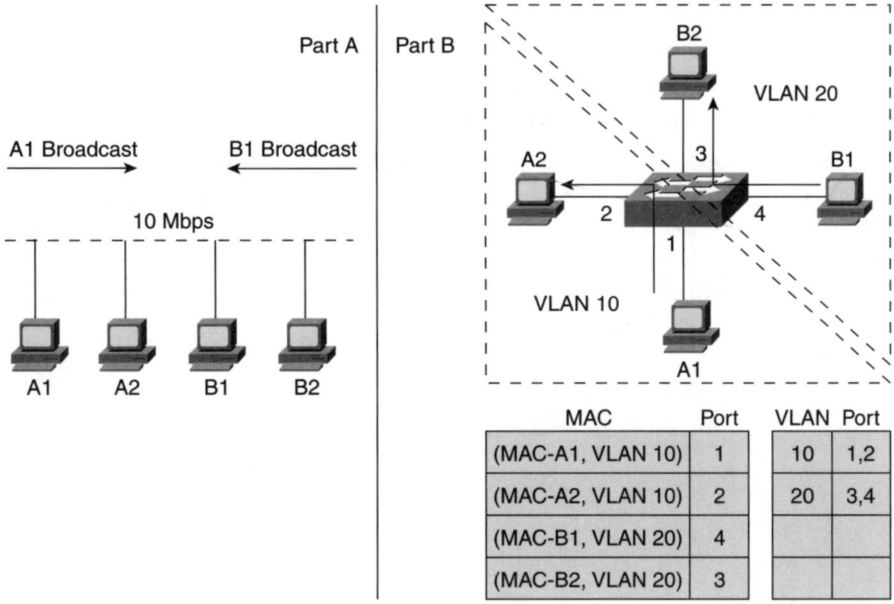

Flooding

If the switch receives a packet with a destination MAC address that does not exist in the bridge table, the switch sends that packet over all its interfaces that belong to the same VLAN assigned to the interface where the packet came in from. The switch does not flood the frame out the port that generated the original frame. This mechanism is called *flooding.* It allows the fast delivery of packets to their destinations even before all MAC addresses have been learned by all switches in the network. The drawback of flooding is that it consumes switch and network resources that otherwise wouldn't have been used if the switch had already learned which port to send the packet to.

VLANs minimize the effect of flooding because they concentrate the flooding within a particular VLAN. The switch uses the VLAN table to map the VLAN number of the port on which the packet arrived to a list of ports that the packet is flooded on.

Using Broadcast and Multicast

Broadcast is used to enable clients to discover resources that are advertised by servers. When a server advertises its services to its clients, it sends broadcast messages to MAC address FFFF FFFF FFFF, which means "all stations." End clients listen to the broadcast and pick up only the broadcasts they are interested in, to minimize their CPU usage. With multicast, a subset of broadcast, a station sends traffic only to a group of stations and not to all stations. Broadcast and multicast addresses are treated as unknown destinations and are flooded over all ports within a VLAN. Some higher-layer protocols such IGMP snooping help mitigate the flooding of IP multicast packets over an L2 switched network by identifying which set of ports a packet is to be flooded on.

Expanding the Network with Trunks

So far you have seen the case of a single L2 switch. An L2 Ethernet-switched network would consist of many interconnected switches with trunk ports. The trunk ports are similar to the access ports used to connect end stations; however, they have the added task of carrying traffic coming in from many VLANs in the network. This scenario is shown in Figure 3-2. Trunk ports could connect Ethernet switches built by different vendors—hence the need for standardization for VLAN tagging mechanisms.

Figure 3-2 *Trunk Ports*

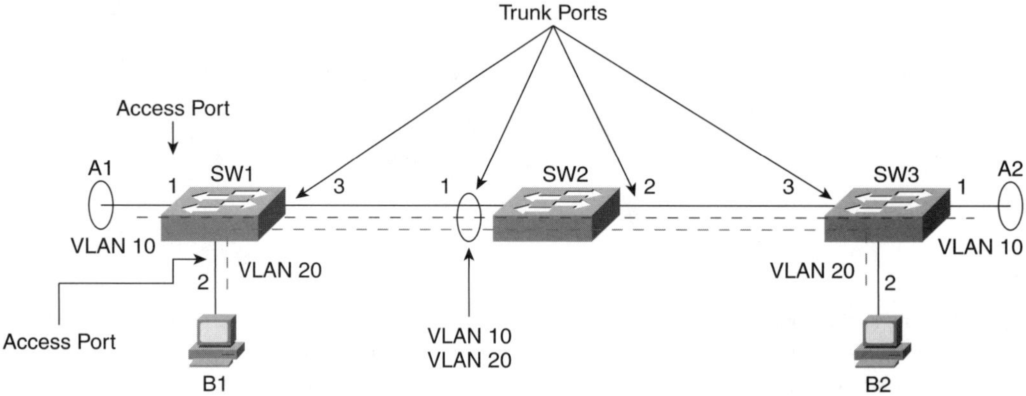

In Figure 3-2, switches SW1 and SW3 have assigned access port 1 with VLAN 10 and access port 2 with VLAN 20. Port 3 is a trunk port that is used to connect to other switches in the network. Note that SW2 in the middle has no access ports and is used only to interconnect trunk ports. You can see that the simplicity of switched Ethernet becomes extremely complex because VLAN assignments need to be tracked inside the network to allow the right traffic to be switched on the right ports. In Frame Relay, ATM, and MPLS, similar complexities do exist, and signaling is introduced to solve the network connectivity issues. Ethernet has *not* defined a signaling protocol. The only mechanisms that Ethernet networks have are third-party applications that surf the network and make it easier to do some VLAN allocations. While these mechanisms work in small enterprise environments, they immediately became showstoppers in larger enterprise deployments and carrier networks. Chapter 4 discusses LDP as a signaling mechanism for delivering Ethernet services. Chapter 7 discusses RSVP-TE and its use in relation to scaling the Ethernet services.

VLAN Tagging

IEEE 802.1Q defines how an Ethernet frame gets tagged with a VLAN ID. The VLAN ID is assigned by the switch and not the end station. The switch assigns a VLAN number to a port, and every packet received on that port gets allocated that VLAN ID. The Ethernet switches switch packets between the same VLANs. Traffic between different VLANs is sent to a routing function within the switch itself (if the switch supports L3 forwarding) or an external router. Figure 3-3 shows how the VLAN tags get inserted inside the untagged VLAN packet.

Figure 3-3 *VLAN Tagged Packet*

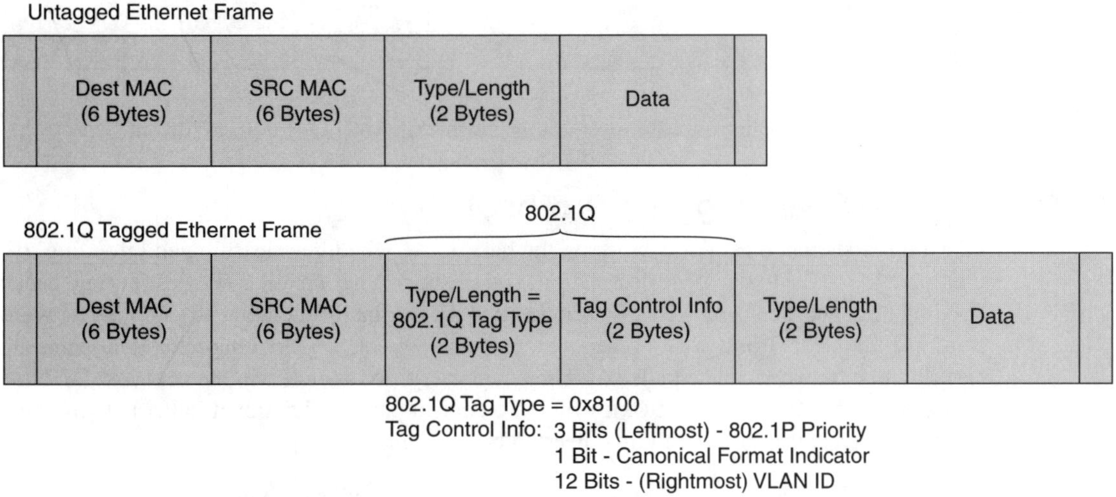

The untagged Ethernet packet consists of the destination MAC address and source MAC address, a Type field, and the data. The 802.1Q tag header gets inserted between the source MAC address and the Type field. It consists of a 2-byte Type field and a 2-byte Tag Control Info field. The 2-byte Type field is set to 0X8100 to indicate an 802.1Q tagged packet. The 2-byte Tag Control Info field consists of the 3 leftmost bits indicating the 802.1P priority and the 12 rightmost bits indicating the VLAN tag ID. The 802.1P field gives the Ethernet packet up to eight different priority levels that can be used to offer different levels of service within the network. The 12-bit VLAN ID field allows the assignment of up to 4096 (2^{12}) VLAN numbers to distinguish the different VLAN tagged packets.

Metro Ethernet applications require extensions to L2 switching that are not defined in the standards. An example is the ability to do VLAN stacking—that is, to do multiple VLAN tagging to the same Ethernet packet and create a stack of VLAN IDs. Different entities can do L2 switching on the different levels of the VLAN stack. Cisco Systems calls this concept *Q-in-Q,* short for *802.1Q-in-802.1Q,* as shown in Figure 3-4.

As shown, an already tagged frame can be tagged again to create a hierarchy. The simplicity of Ethernet, the lack of standardization for many such extensions, the reliance on STP, and the explosion of MAC addresses contribute to the lack of confidence of many providers in deploying a large-scale, all-Ethernet network.

Figure 3-4 *Q-in-Q*

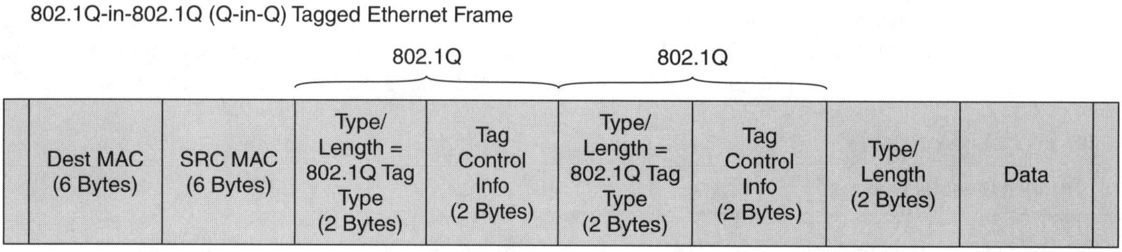

VLAN tag support is discussed more in the section "VLAN Tag Support Attribute."

The Need for the Spanning Tree Protocol

L2 Ethernet-switched networks work on the basis of MAC address learning and flooding. If multiple paths exist to the same destination, and the packet has an unknown destination, packet flooding might cause the packet to be sent back to the original switch that put it on the network, causing a broadcast storm. STP prevents loops in the network by blocking redundant paths and ensuring that only one active path exists between every two switches in the network. STP uses bridge protocol data units (BPDUs), which are control packets that travel in the network and identify which path, and hence ports, need to be blocked.

The next section covers in detail the Ethernet services concepts as defined by the Metro Ethernet Forum.

Metro Ethernet Services Concepts

The Metro Ethernet Forum is a nonprofit organization that has been active in defining the scope, concepts, and terminology for deploying Ethernet services in the metro. Other standards bodies, such as the Internet Engineering Task Force (IETF), have also defined ways of scaling Ethernet services through the use of MPLS. While the terminologies might differ slightly, the concepts and directions taken by these different bodies are converging.

For Ethernet services, the MEF defines a set of attributes and parameters that describe the service and SLA that are set between the metro carrier and its customer.

Ethernet Service Definition

The MEF defines a User-to-Network Interface (UNI) and Ethernet Virtual Connection (EVC). The UNI is a standard Ethernet interface that is the point of demarcation between the customer equipment and the service provider's metro Ethernet network.

The EVC is defined by the MEF as "an association of two or more UNIs." In other words, the EVC is a logical tunnel that connects two (P2P) or more (MP2MP) sites, enabling the transfer of Ethernet frames between them. The EVC also acts as a separation between the different customers and provides data privacy and security similar to Frame Relay or ATM permanent virtual circuits (PVCs).

The MEF has defined two Ethernet service types:

- **Ethernet Line Service (ELS)**—This is basically a point-to-point (P2P) Ethernet service.
- **Ethernet LAN Service (E-LAN)**—This is a multipoint-to-multipoint (MP2MP) Ethernet service.

The Ethernet Line Service provides a P2P EVC between two subscribers, similar to a Frame Relay or private leased-line service (see Figure 3-5).

Figure 3-5 *Ethernet Service Concepts*

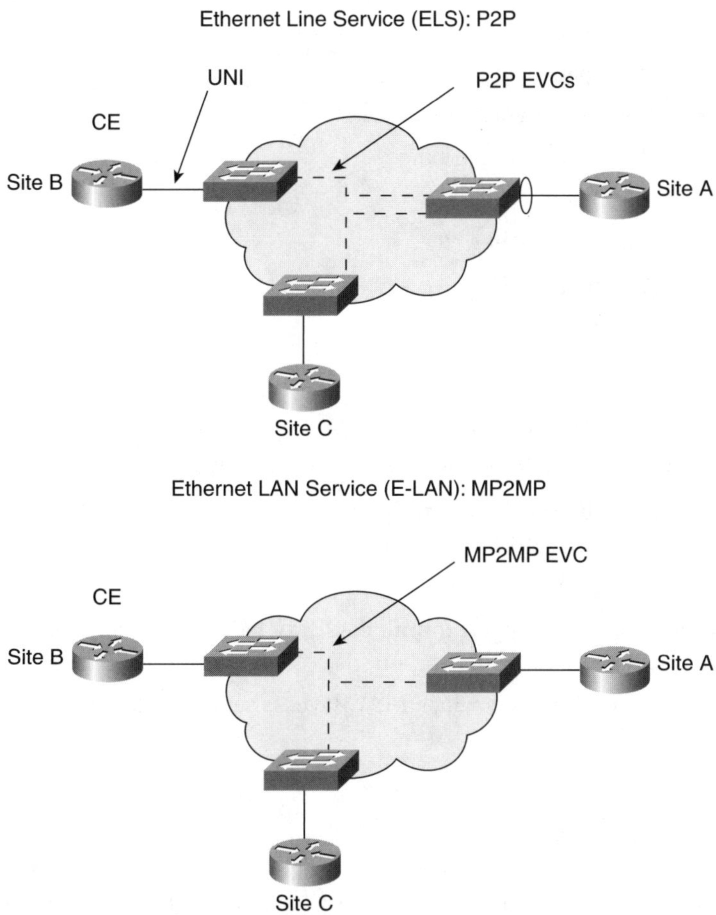

Figure 3-5 also illustrates the E-LAN, which provides multipoint connectivity between multiple subscribers in exactly the same manner as an Ethernet-switched network. An E-LAN service offers the most flexibility in providing a VPN service because one EVC touches all sites. If a new site is added to the VPN, the new site participates in the EVC and has automatic connectivity to all other sites.

Ethernet Service Attributes and Parameters

The MEF has developed an Ethernet services framework to help subscribers and service providers have a common nomenclature when talking about the different service types and their attributes. For each of the two service types, ELS and E-LAN, the MEF has defined the following service attributes and their corresponding parameters that define the capabilities of the service type:

- Ethernet physical interface attribute
- Traffic parameters
- Performance parameters
- Class of service parameters
- Service frame delivery attribute
- VLAN tag support attribute
- Service multiplexing attribute
- Bundling attribute
- Security filters attribute

Ethernet Physical Interface Attribute

The Ethernet physical interface attribute has the following parameters:

- **Physical medium**—Defines the physical medium per the IEEE 802.3 standard. Examples are 10BASE-T, 100BASE-T, and 1000BASE-X.
- **Speed**—Defines the Ethernet speed: 10 Mbps, 100 Mbps, 1 Gbps, or 10 Gbps.
- **Mode**—Indicates support for full duplex or half duplex and support for autospeed negotiation between Ethernet ports.
- **MAC layer**—Specifies which MAC layer is supported as specified in the 802.3-2002 standard.

Traffic Parameters

The MEF has defined a set of bandwidth profiles that can be applied at the UNI or to an EVC. A bandwidth profile is a limit on the rate at which Ethernet frames can traverse the UNI or the

EVC. Administering the bandwidth profiles can be a tricky business. For P2P connections where there is a single EVC between two sites, it is easy to calculate a bandwidth profile coming in and out of the pipe. However, for the cases where a multipoint service is delivered and there is the possibility of having multiple EVCs on the same physical interface, it becomes difficult to determine the bandwidth profile of an EVC. In such cases, limiting the bandwidth profile per UNI might be more practical.

The Bandwidth Profile service attributes are as follows:

- Ingress and egress bandwidth profile per UNI
- Ingress and egress bandwidth profile per EVC
- Ingress and egress bandwidth profile per CoS identifier
- Ingress bandwidth profile per destination UNI per EVC
- Egress bandwidth profile per source UNI per EVC

The Bandwidth Profile service attributes consist of the following traffic parameters:

- **CIR (Committed Information Rate)**—This is the minimum guaranteed throughput that the network must deliver for the service under normal operating conditions. A service can support a CIR per VLAN on the UNI interface; however, the sum of all CIRs should not exceed the physical port speed. The CIR has an additional parameter associated with it called the Committed Burst Size (CBS). The CBS is the size up to which subscriber traffic is allowed to burst in profile and not be discarded or shaped. The in-profile frames are those that meet the CIR and CBS parameters. The CBS may be specified in KB or MB. If, for example, a subscriber is allocated a 3-Mbps CIR and a 500-KB CBS, the subscriber is guaranteed a minimum of 3 Mbps and can burst up to 500 KB of traffic and still remain within the SLA limits. If the traffic bursts above 500 KB, the traffic may be dropped or delayed.
- **PIR (Peak Information Rate)**—The PIR specifies the rate above the CIR at which traffic is allowed into the network and that may get delivered if the network is not congested. The PIR has an additional parameter associated with it called the Maximum Burst Size (MBS). The MBS is the size up to which the traffic is allowed to burst without being discarded. The MBS can be specified in KB or MB, similar to CBS. A sample service may provide a 3-Mbps CIR, 500-KB CBS, 10-Mbps PIR, and 1-MB MBS. In this case, the following behavior occurs:
  - Traffic is less than or equal to CIR (3 Mbps)—Traffic is in profile with a guaranteed delivery. Traffic is also in profile if it bursts to CBS (500 KB) and may be dropped or delayed if it bursts beyond 500 KB.
  - Traffic is more than CIR (3 Mbps) and less than PIR (10 Mbps)—Traffic is out of profile. It may get delivered if the network is not congested and the burst size is less than MBS (1 MB).
  - Traffic is more than PIR (10 Mbps)—Traffic is discarded.

Performance Parameters

The performance parameters indicate the service quality experienced by the subscriber. They consist of the following:

- Availability
- Delay
- Jitter
- Loss

Availability

Availability is specified by the following service attributes:

- **UNI Service Activation Time**—Specifies the time from when the new or modified service order is placed to the time service is activated and usable. Remember that the main value proposition that an Ethernet service claims is the ability to cut down the service activation time to hours versus months with respect to the traditional telco model.
- **UNI Mean Time to Restore (MTTR)**—Specifies the time it takes from when the UNI is unavailable to when it is restored. Unavailability can be caused by a failure such as a fiber cut.
- **EVC Service Activation Time**—Specifies the time from when a new or modified service order is placed to when the service is activated and usable. The EVC service activation time begins when all UNIs are activated. For a multipoint EVC, for example, the service is considered active when all UNIs are active and operational.
- **EVC Availability**—Specifies how often the subscriber's EVC meets or exceeds the delay, loss, and jitter service performance over the same measurement interval. If an EVC does not meet the performance criteria, it is considered unavailable.
- **EVC (MTTR)**—Specifies the time from when the EVC is unavailable to when it becomes available again. Many restoration mechanisms can be used on the physical layer (L1), the MAC layer (L2), or the network layer (L3).

Delay

Delay is a critical parameter that significantly impacts the quality of service (QoS) for real-time applications. Delay has traditionally been specified in one direction as one-way delay or end-to-end delay. The delay between two sites in the metro is an accumulation of delays, starting from one UNI at one end, going through the metro network, and going through the UNI on the other end. The delay at the UNI is affected by the line rate at the UNI connection and the supported Ethernet frame size. For example, a UNI connection with 10 Mbps and 1518-byte frame size would cause 1.2 milliseconds (ms) of transmission delay ($1518 * 8 / 10^6$).

The metro network itself introduces additional delays based on the network backbone speed and level of congestion. The delay performance is defined by the 95th percentile (95 percent) of the delay of successfully delivered egress frames over a time interval. For example, a delay of 15 ms over 24 hours means that over a period of 24 hours, 95 percent of the "delivered" frames had a one-way delay of less than or equal to 15 ms.

The delay parameter is used in the following attributes:

- Ingress and egress bandwidth profile per CoS identifier (UNI service attribute)
- Class of service (EVC service attribute)

Jitter

Jitter is another parameter that affects the service quality. Jitter is also known as delay variation. Jitter has a very adverse effect on real-time applications such as IP telephony. The jitter parameter is used in the following service attributes:

- Ingress and egress bandwidth profile per CoS identifier (UNI service attribute)
- Class of service (EVC service attribute)

Loss

Loss indicates the percentage of Ethernet frames that are in-profile and that are not reliably delivered between UNIs over a time interval. On a P2P EVC, for example, if 100 frames have been sent from a UNI on one end and 90 frames that are in profile have been received on the other end, the loss would be (100 – 90) / 100 = 10%. Loss can have adverse effects, depending on the application. Applications such as e-mail and HTTP web browser requests can tolerate more loss than VoIP, for example. The loss parameter is used in the following attributes:

- Ingress and egress bandwidth profile per CoS identifier (UNI service attribute)
- Class of service (EVC service attribute)

Class of Service Parameters

Class of service (CoS) parameters can be defined for metro Ethernet subscribers based on various CoS identifiers, such as the following:

- **Physical port**—This is the simplest form of QoS that applies to the physical port of the UNI connection. All traffic that enters and exits the port receives the same CoS.
- **Source/destination MAC addresses**—This type of classification is used to give different types of service based on combinations of source and destination MAC addresses. While this model is very flexible, it is difficult to administer, depending on the service itself. If the customer premises equipment (CPE) at the ends of the connections are Layer 2 switches that are part of a LAN-to-LAN service, hundreds or thousands of MAC addresses might have to be monitored. On the other hand, if the CPEs are routers, the MAC addresses that are monitored are those of the router interfaces themselves. Hence, the MAC addresses are much more manageable.

- **VLAN ID**—This is a very practical way of assigning CoS if the subscriber has different services on the physical port where a service is defined by a VLAN ID (these would be the carrier-assigned VLANs).
- **802.1p value**—The 802.1p field allows the carrier to assign up to eight different levels of priorities to the customer traffic. Ethernet switches use this field to specify some basic forwarding priorities, such as that frames with priority number 7 get forwarded ahead of frames with priority number 6, and so on. This is one method that can be used to differentiate between VoIP traffic and regular traffic or between high-priority and best-effort traffic. In all practicality, service providers are unlikely to exceed two or three levels of priority, for the sake of manageability.
- **Diffserv/IP ToS**—The IP ToS field is a 3-bit field inside the IP packet that is used to provide eight different classes of service known as IP precedence. This field is similar to the 802.1p field if used for basic forwarding priorities; however, it is located inside the IP header rather than the Ethernet 802.1Q tag header. Diffserv has defined a more sophisticated CoS scheme than the simple forwarding priority scheme defined by ToS. Diffserv allows for 64 different CoS values, called Diffserv codepoints (DSCPs). Diffserv includes different per-hop behaviors (PHBs), such as Expedited Forwarding (EF) for a low delay, low-loss service, four classes of Assured Forwarding (AF) for bursty real-time and non-real-time services, Class Selector (CS) for some backward compatibility with IP ToS, and Default Forwarding (DF) for best-effort services.

  Although Diffserv gives much more flexibility to configure CoS parameters, service providers are still constrained with the issue of manageability. This is similar to the airline QoS model. Although there are so many ways to arrange seats and who sits where and so many types of food service and luggage service to offer travelers, airlines can manage at most only three or four levels of service, such as economy, economy plus, business class, and first class. Beyond that, the overhead of maintaining these services and the SLAs associated with them becomes cost-prohibitive.

Service Frame Delivery Attribute

Because the metro network behaves like a switched LAN, you must understand which frames need to flow over the network and which do not. On a typical LAN, the frames traversing the network could be data frames or control frames. Some Ethernet services support delivery of all types of Ethernet protocol data units (PDUs); others may not. To ensure the full functionality of the subscriber network, it is important to have an agreement between the subscriber and the metro carriers on which frames get carried. The EVC service attribute can define whether a particular frame is discarded, delivered unconditionally, or delivered conditionally for each ordered UNI pair. The different possibilities of the Ethernet data frames are as follows:

- **Unicast frames**—These are frames that have a specified destination MAC address. If the destination MAC address is known by the network, the frame gets delivered to the exact destination. If the MAC address is unknown, the LAN behavior is to flood the frame within the particular VLAN.

- **Multicast frames**—These are frames that are transmitted to a select group of destinations. This would be any frame with the least significant bit (LSB) of the destination address set to 1, except for broadcast, where all bits of the MAC destination address are set to 1.
- **Broadcast frames**—IEEE 802.3 defines the broadcast address as a destination MAC address, FF-FF-FF-FF-FF-FF.

Layer 2 Control Processing packets are the different L2 control-protocol packets needed for specific applications. For example, BPDU packets are needed for STP. The provider might decide to tunnel or discard these packets over the EVC, depending on the service. The following is a list of currently standardized L2 protocols that can flow over an EVC:

- **IEEE 802.3x MAC control frames**—802.3.x is an XON/XOFF flow-control mechanism that lets an Ethernet interface send a PAUSE frame in case of traffic congestion on the egress of the Ethernet switch. The 802.3x MAC control frames have destination address 01-80-C2-00-00-01.
- **Link Aggregation Control Protocol (LACP)**—This protocol allows the dynamic bundling of multiple Ethernet interfaces between two switches to form an aggregate bigger pipe. The destination MAC address for these control frames is 01-80-C2-00-00-02.
- **IEEE 802.1x port authentication**—This protocol allows a user (an Ethernet port) to be authenticated into the network via a back-end server, such as a RADIUS server. The destination MAC address is 01-80-C2-00-00-03.
- **Generic Attribute Registration Protocol (GARP)**—The destination MAC address is 01-80-C2-00-00-2X.
- **STP**—The destination MAC address is 01-80-C2-00-00-00.
- **All-bridge multicast**—The destination MAC address is 01-80-C2-00-00-10.

VLAN Tag Support Attribute

VLAN tag support provides another set of capabilities that are important for service frame delivery. Enterprise LANs are single-customer environments, meaning that the end users belong to a single organization. VLAN tags within an organization are indicative of different logical broadcast domains, such as different workgroups. Metro Ethernet creates a different environment in which the Ethernet network supports multiple enterprise networks that share the same infrastructure, and in which each enterprise network can still have its own segmentation. Support for different levels of VLANs and the ability to manipulate VLAN tags become very important.

Consider the example of an MTU building in which the metro provider installs a switch in the basement that offers multiple Ethernet connections to different small offices in the building. In this case, from a carrier perspective, each customer is identified by the physical Ethernet interface port that the customer connects to. This is shown in Figure 3-6.

Although identifying the customer itself is easy, isolating the traffic between the different customers becomes an interesting issue and requires some attention on the provider's part. Without special attention, traffic might get exchanged between the different customers in the building through the basement switch. You have already seen in the section "L2 Switching Basics" that VLANs can be used to separate physical segments into many logical segments; however, this works in a single-customer environment, where the VLAN has a global meaning. In a multicustomer environment, each customer can have its own set of VLANs that overlap with VLANs from another customer. To work in this environment, carriers are adopting a model very similar to how Frame Relay and ATM services have been deployed. In essence, each customer is given service identifiers similar to Frame Relay data-link connection identifiers (DLCIs), which identify EVCs over which the customer's traffic travels. In the case of Ethernet, the VLAN ID given by a carrier becomes that identifier. This is illustrated in Figure 3-7.

Figure 3-6 *Ethernet in Multicustomer Environments*

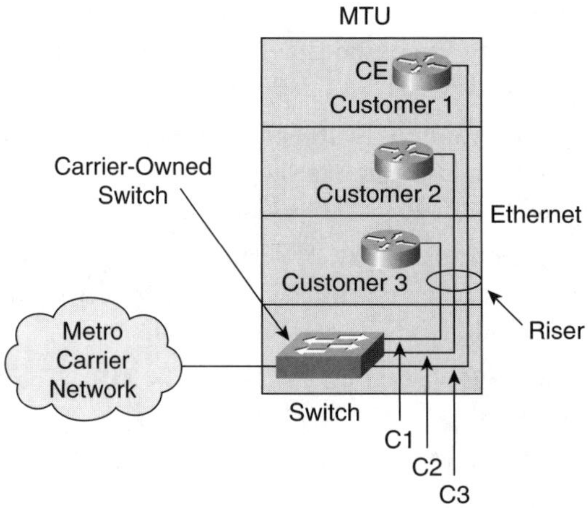

In this example, the carrier needs to assign to each physical port a set of VLAN IDs that are representative of the services sold to each customer. Customer 1, for example, is assigned VLAN 10, customer 2 is assigned VLAN 20, and customer 3 is assigned VLAN 30. VLANs 10, 20, and 30 are carrier-assigned VLANs that are independent of the customer's internal VLAN assignments. To make that distinction, the MEF has given the name CE-VLANs to the customer-internal VLANs. The customers themselves can have existing VLAN assignments (CE-VLANs) that overlap with each other and the carrier's VLAN. There are two types of VLAN tag support:

- VLAN Tag Preservation/Stacking
- VLAN Tag Translation/Swapping

VLAN Tag Preservation/Stacking

With VLAN Tag Preservation, all Ethernet frames received from the subscriber need to be carried untouched within the provider's network across the EVC. This means that the VLAN ID at the ingress of the EVC is equal to the VLAN ID on the egress. This is typical of services such as LAN extension, where the same LAN is extended between two different locations and the enterprise-internal VLAN assignments need to be preserved. Because the carrier's Ethernet switch supports multiple customers with overlapping CE-VLANs, the carrier's switch needs to be able to stack its own VLAN assignment on top of the customer's VLAN assignment to keep the separation between the traffic of different customers. This concept is called 802.1Q-in-802.1Q or Q-in-Q stacking, as explained earlier in the section "VLAN Tagging." With Q-in-Q, the carrier VLAN ID becomes indicative of the EVC, while the customer VLAN ID (CE-VLAN) is indicative of the internals of the customer network and is hidden from the carrier's network.

Figure 3-7 *Logical Separation of Traffic and Services*

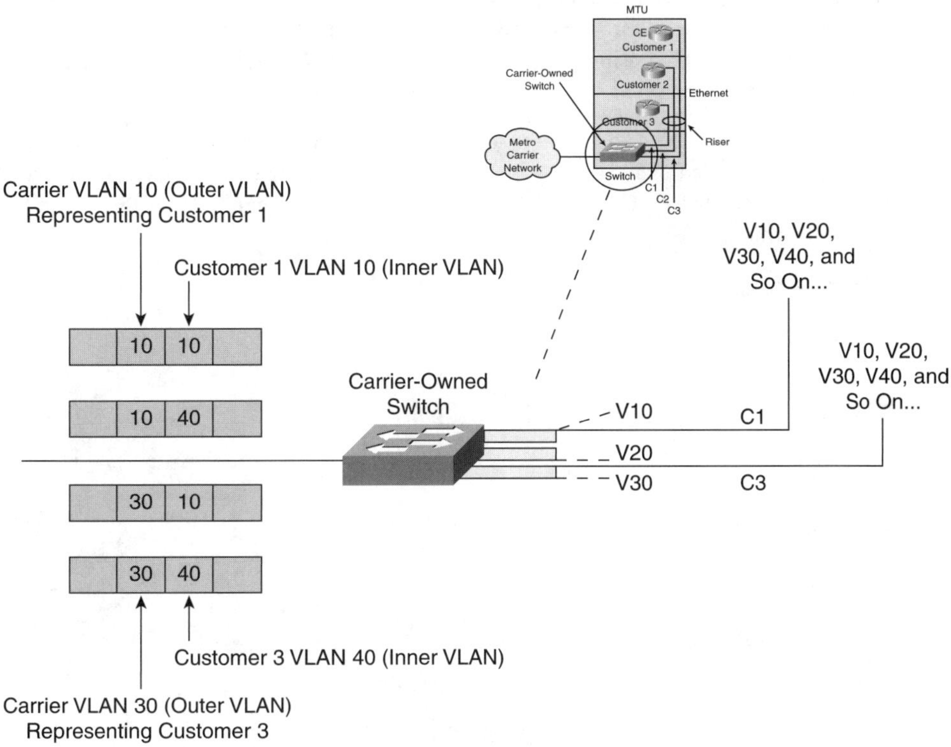

WARNING The Q-in-Q function is not standardized, and many vendors have their own variations. For the service to work, the Q-in-Q function must work on a "per-port" basis, meaning that each customer can be tagged with a different carrier VLAN tag. Some enterprise switches on the market can perform a double-tagging function; however, these switches can assign only a single VLAN-ID as a carrier ID for the whole switch. These types of switches work only if a single customer is serviced and the carrier wants to be able to carry the customer VLANs transparently within its network. These switches do not work when the carrier switch is servicing multiple customers, because it is impossible to differentiate between these customers using a single carrier VLAN tag.

VLAN Tag Translation/Swapping

VLAN Tag Translation or Swapping occurs when the VLAN tags are local to the UNI, meaning that the VLAN tag value, if it exists on one side of the EVC, is independent of the VLAN tag values on the other side. In the case where one side of the EVC supports VLAN tagging and the other side doesn't, the carrier removes the VLAN tag from the Ethernet frames before they are delivered to the destination.

Another case is two organizations that have merged and want to tie their LANs together, but the internal VLAN assignments of each organization do not match. The provider can offer a service where the VLANs are removed from one side of the EVC and are translated to the correct VLANs on the other side of the EVC. Without this service, the only way to join the two organizations is via IP routing, which ignores the VLAN assignments and delivers the traffic based on IP addresses.

Another example of tag translation is a scenario where different customers are given Internet connectivity to an ISP. The carrier gives each customer a separate EVC. The carrier assigns its own VLAN-ID to the EVC and strips the VLAN tag before handing off the traffic to the ISP. This is illustrated in Figure 3-8.

Figure 3-8 *VLAN Translation*

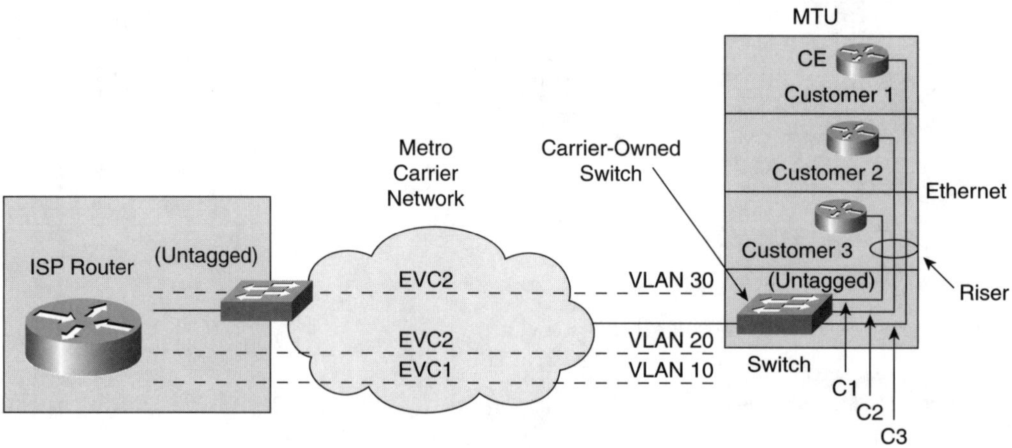

Figure 3-8 shows the metro carrier delivering Internet connectivity to three customers. The carrier is receiving untagged frames from the CPE routers located at each customer premises. The carrier inserts a VLAN tag 10 for all of customer 1's traffic, VLAN 20 for customer 2's traffic, and VLAN 30 for customer 3's traffic. The carrier uses the VLAN tags to separate the three customers' traffic within its own network. At the point of presence (POP), the VLAN tags are removed from all EVCs and handed off to an ISP router, which is offering the Internet IP service.

Service Multiplexing Attribute

Service multiplexing is used to support multiple instances of EVCs on the same physical connection. This allows the same customer to have different services with the same Ethernet wire.

Bundling Attribute

The Bundling service attribute enables two or more VLAN IDs to be mapped to a single EVC at a UNI. With bundling, the provider and subscriber must agree on the VLAN IDs used at the UNI and the mapping between each VLAN ID and a specific EVC. A special case of bundling is where every VLAN ID at the UNI interface maps to a single EVC. This service attribute is called *all-to-one bundling.*

Security Filters Attribute

Security filters are MAC access lists that the carrier uses to block certain addresses from flowing over the EVC. This could be an additional service the carrier can offer at the request of the subscriber who would like a level of protection against certain MAC addresses. MAC addresses that match a certain access list could be dropped or allowed.

Tables 3-1 and 3-2 summarize the Ethernet service attributes and their associated parameters for UNI and EVCs.

Table 3-1 *UNI Service Attributes*

UNI Service Attribute	Parameter Values or Range of Values
Physical medium	A standard Ethernet physical interface.
Speed	10 Mbps, 100 Mbps, 1 Gbps, or 10 Gbps.
Mode	Full-duplex or autospeed negotiation.
MAC layer	Ethernet and/or IEEE 802.3-2002.
Service multiplexing	Yes or no. If yes, all-to-one bundling must be no.
Bundling	Yes or no. Must be no if all-to-one bundling is yes and yes if all-to-one bundling is no.
All-to-one bundling	Yes or no. If yes, service multiplexing and bundling must be no. Must be no if bundling is yes.
Ingress and egress bandwidth profile per UNI	No or one of the following parameters: CIR, CBS, PIR, MBS. If no, no bandwidth profile per UNI is set; otherwise, the traffic parameters CIR, CBS, PIR, and MBS need to be set.

continues

Table 3-1 *UNI Service Attributes (Continued)*

UNI Service Attribute	Parameter Values or Range of Values
Ingress and egress bandwidth profile per EVC	No or one of the following parameters: CIR, CBS, PIR, MBS.
Ingress and egress bandwidth profile per CoS identifier	No or one of the following parameters: CIR, CBS, PIR, MBS. If one of the parameters is chosen, specify the CoS identifier, Delay value, Jitter value, Loss value. If no, no bandwidth profile per CoS identifier is set; otherwise, the traffic parameters CIR, CBS, PIR, and MBS need to be set.
Ingress and egress bandwidth profile per destination UNI per EVC	No or one of the following parameters: CIR, CBS, PIR, MBS.
Egress bandwidth profile per source UNI per EVC	No or one of the following parameters: CIR, CBS, PIR, MBS.
Layer 2 Control Protocol processing	Process, discard, or pass to EVC the following control protocol frames: • IEEE 802.3x MAC control • Link Aggregation Control Protocol (LACP) • IEEE 802.1x port authentication • Generic Attribute Registration Protocol (GARP) • STP • Protocols multicast to all bridges in a bridged LAN
UNI service activation time	Time value

Table 3-2 *EVC Service Attributes*

EVC Service Attribute	Type of Parameter Value
EVC Type	P2P or MP2MP
CE-VLAN ID preservation	Yes or no
CE-VLAN CoS preservation	Yes or no
Unicast frame delivery	Discard, deliver unconditionally, or deliver conditionally for each ordered UNI pair
Multicast frame delivery	Discard, deliver unconditionally, or deliver conditionally for each ordered UNI pair
Broadcast frame delivery	Discard, deliver unconditionally, or deliver conditionally for each ordered UNI pair

Table 3-2 *EVC Service Attributes (Continued)*

EVC Service Attribute	Type of Parameter Value
Layer 2 Control Protocol processing	Discard or tunnel the following control frames: • IEEE 802.3x MAC control • Link Aggregation Control Protocol (LACP) • IEEE 802.1x port authentication • Generic Attribute Registration Protocol (GARP) • STP • Protocols multicast to all bridges in a bridged LAN
EVC service activation time	Time value
EVC availability	Time value
EVC mean time to restore	Time value
Class of service	CoS identifier, Delay value, Jitter value, Loss value This assigns the Class of Service Identifier to the EVC

Example of an L2 Metro Ethernet Service

This section gives an example of an L2 metro Ethernet service and how all the parameters defined by the MEF are applied. The example attempts to highlight many of the definitions and concepts discussed in this chapter.

If you have noticed, the concept of VPNs is inherent in L2 Ethernet switching. The carrier VLAN is actually a VPN, and all customer sites within the same carrier VLAN form their own user group and exchange traffic independent of other customers on separate VLANs.

The issue of security arises in dealing with VLAN isolation between customers; however, because the metro network is owned by a central entity (such as the metro carrier), security is enforced. First of all, the access switches in the customer basement are owned and administered by the carrier, so physical access is prevented. Second, the VLANs that are switched in the network are assigned by the carrier, so VLAN isolation is guaranteed. Of course, misconfiguration of switches and VLAN IDs could cause traffic to be mixed, but this problem can occur with any technology used, not just Ethernet. Issues of security always arise in public networks whether they are Ethernet, IP, MPLS, or Frame Relay networks. The only definite measure to ensure security is to have the customer-to-customer traffic encrypted at the customer sites and to have the customers administer that encryption.

Figure 3-9 shows an example of an L2 metro Ethernet VPN. This example attempts to show in a practical way how many of the parameters and the concepts that are discussed in this chapter are used.

Figure 3-9 *All-Ethernet L2 Metro Service Example*

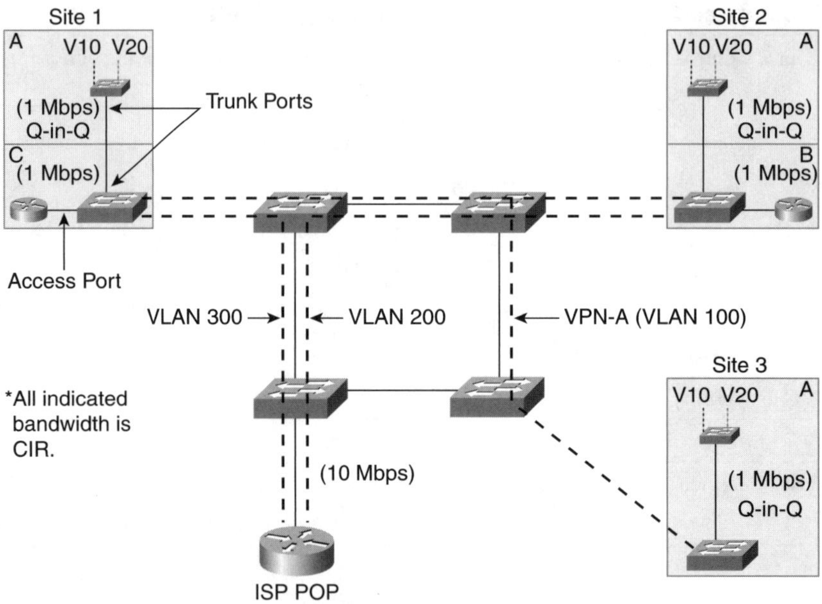

Figure 3-9 shows a metro carrier offering an L2 MP2MP VPN service to customer A and a packet leased-line service (comparable to a traditional T1 leased line) to an ISP. In turn, the ISP is offering Internet service to customers B and C. It is assumed that customer A connects to the carrier via L2 Ethernet switches and customers B and C connect via IP routers. Notice the difference between access ports and trunk ports on the Ethernet switches. The ports connecting the customer's Ethernet switch to the carrier's Ethernet switch are trunk ports, because these ports are carrying multiple VLANs between the two switches. When the carrier's switch port is configured for Q-in-Q, it encapsulates the customers' CE-VLAN tags VLAN 10 and VLAN 20 inside the carrier VLAN 100. On the other hand, the ports connecting the customer router with the carrier switch are access ports and are carrying untagged traffic from the router. Tables 3-3 and 3-4 describe the UNI and EVC service attributes for customers A, B, and C as defined by the MEF.

Table 3-3 *Customer A E-LAN UNI Service Attributes*

Customer A E-LAN UNI Service Attribute	Parameter Values or Range of Values
Physical medium	Standard Ethernet physical interfaces
Speed	100 Mbps site 1, 10 Mbps sites 2 and 3
Mode	Full duplex all sites
MAC layer	IEEE 802.3-2002
Service multiplexing	No

Table 3-3 *Customer A E-LAN UNI Service Attributes (Continued)*

Customer A E-LAN UNI Service Attribute	Parameter Values or Range of Values
Bundling	No
All-to-one bundling	Yes
Ingress and egress bandwidth profile per CoS identifier	All sites CoS 1: • CIR = 1 Mbps, CBS = 100 KB, PIR = 2 Mbps, MBS = 100 KB • CoS ID = 802.1p 6–7 • Delay < 10 ms, Loss < 1% All sites CoS 0: • CIR = 1 Mbps, CBS = 100 KB, PIR = 10 Mbps, MBS = 100 KB • CoS ID = 802.1p 0–5, Delay < 35 ms, Loss < 2%
Layer 2 Control Protocol processing	• Process IEEE 802.3x MAC control • Process Link Aggregation Control Protocol (LACP) • Process IEEE 802.1x port authentication • Pass Generic Attribute Registration Protocol (GARP) • Pass STP • Pass protocols multicast to all bridges in a bridged LAN
UNI service activation time	One hour after equipment is installed

Note in Table 3-3 that customer A is given only one MP2P EVC; hence, there is no service multiplexing. All customer VLANs 10 and 20 are mapped to the MP2MP EVC in the form of carrier VLAN 100. Customer A is given two Class of Service profiles—CoS 1 and CoS 0. Each profile has its set of performance attributes. Profile 1, for example, is applied to high-priority traffic, as indicated by 802.1p priority levels 6 and 7. Profile 0 is lower priority, with less-stringent performance parameters. For customer A, the metro carrier processes the 802.3x and LACP frames on the UNI connection and passes other L2 control traffic that belongs to the customer. Passing the STP control packets, for example, prevents any potential loops within the customer network, in case the customer has any L2 backdoor direct connection between its different sites.

Table 3-4 *Customer A E-LAN EVC Service Attributes*

Customer A E-LAN EVC Service Attribute	Type of Parameter Value
EVC type	MP2MP
CE-VLAN ID preservation	Yes

continues

Table 3-4 *Customer A E-LAN EVC Service Attributes (Continued)*

Customer A E-LAN EVC Service Attribute	Type of Parameter Value
CE-VLAN CoS preservation	Yes
Unicast frame delivery	Deliver unconditionally for each UNI pair
Multicast frame delivery	Deliver unconditionally for each UNI pair
Broadcast frame delivery	Deliver unconditionally for each UNI pair
Layer 2 Control Protocol processing	Tunnel the following control frames: • IEEE 802.3x MAC control • Link Aggregation Control Protocol (LACP) • IEEE 802.1x port authentication • Generic Attribute Registration Protocol (GARP) • STP • Protocols multicast to all bridges in a bridged LAN
EVC service activation time	Twenty minutes after UNI is operational
EVC availability	Three hours
EVC mean time to restore	One hour
Class of service	All sites CoS 1: • CoS ID = 802.1p 6–7 • Delay < 10 ms, Loss < 1%, Jitter (value) All sites CoS 0: • CoS ID = 802.1p 0–5, Delay < 35 ms, Loss < 2%, Jitter (value)

The EVC service parameters for customer A indicate that the EVC is an MP2MP connection and the carrier transparently moves the customer VLANs between sites. The carrier does this using Q-in-Q tag stacking with a carrier VLAN ID of 100. The carrier also makes sure that the 802.1p priority fields that the customer sends are still carried within the network. Note that the carrier allocates priority within its network whichever way it wants as long as the carrier delivers the SLA agreed upon with the customer as described in the CoS profiles. For customer A, the carrier passes all unicast, multicast, and broadcast traffic and also tunnels all L2 protocols between the different sites.

Tables 3-5 and 3-6 describe customers B and C and ISP POP service profile for the Internet connectivity service. These are the service attributes and associated parameters for customers

B and C as well as the service attributes and associated parameters for the ISP POP offering Internet connectivity to these customers.

Table 3-5 *Customers B and C and ISP POP UNI Service Attributes*

Customers B and C and ISP POP Internet Access UNI Service Attribute	Parameter Values or Range of Values
Physical medium	Standard Ethernet physical interfaces
Speed	10 Mbps for customers B and C, 100 Mbps for the ISP POP
Mode	Full duplex all sites
MAC layer	IEEE 802.3-2002
Service multiplexing	Yes, only at ISP POP UNI
Bundling	No
All-to-one bundling	No
Ingress and egress bandwidth profile per EVC	Customers B and C CIR = 1 Mbps, CBS = 100 KB, PIR = 2 Mbps, MBS = 100 KB ISP POP CIR = 10 Mbps, CBS = 1 MB, PIR = 100 Mbps, MBS = 1 MB
Layer 2 Control Protocol processing	Discard the following control frames: • IEEE 802.3x MAC control • Link Aggregation Control Protocol (LACP) • IEEE 802.1x port authentication • Generic Attribute Registration Protocol (GARP) • STP • Protocols multicast to all bridges in a bridged LAN
UNI service activation time	One hour after equipment is installed

For customers B and C and ISP POP UNI service parameters, because two different P2P EVCs (carrier VLANs 200 and 300) are configured between the customers and the ISP POP, service multiplexing occurs at the ISP UNI connection where two EVCs are multiplexed on the same physical connection. For this Internet access scenario, routers are the customer premises equipment, so it is unlikely that the customer will send any L2 control-protocol packets to the carrier. In any case, all L2 control-protocol packets are discarded if any occur.

Table 3-6 *Customers B and C and ISP POP EVC Service Attributes*

Customers B and C and ISP POP Internet Access EVC Service Attribute	Type of Parameter Value
EVC type	P2P
CE-VLAN ID preservation	No; mapped VLAN ID for provider use
CE-VLAN CoS preservation	No
Unicast frame delivery	Deliver unconditionally for each UNI pair
Multicast frame delivery	Deliver unconditionally for each UNI pair
Broadcast frame delivery	Deliver unconditionally for each UNI pair
Layer 2 Control Protocol processing	Discard the following control frames: • IEEE 802.3x MAC control • Link Aggregation Control Protocol (LACP) • IEEE 802.1x port authentication • Generic Attribute Registration Protocol (GARP) • STP • Protocols multicast to all bridges in a bridged LAN
EVC service activation time	Twenty minutes after UNI is operational
EVC availability	Three hours
EVC mean time to restore	One hour
Class of service	One CoS service is supported: Delay < 30 ms, Loss < 1%, Jitter (value)

The EVC parameters indicate that the carrier is not preserving any customer VLANs or CoS info. Also, because this is an Internet access service, normally the provider receives untagged frames from the CPE router. The provider can map those frames to carrier VLANs 200 and 300 if it needs to separate the traffic in its network. The VLAN IDs are normally stripped off before given to the ISP router.

Challenges with All-Ethernet Metro Networks

All-Ethernet metro networks pose many scalability and reliability challenges. The following are some of the issues that arise with an all-Ethernet control plane:

- Restrictions on the number of customers
- Service monitoring
- Scaling the L2 backbone

- Service provisioning
- Interworking with legacy deployments

The following sections describe each of these challenges.

Restrictions on the Number of Customers

The Ethernet control plane restricts the carrier to 4096 customers, because the 802.1Q defines 12 bits that can be used as a VLAN ID, which restricts the number of VLANs to $2^{12} = 4096$. Remember that although Q-in-Q allows the customer VLANs (CE-VLANs) to be hidden from the carrier network, the carrier is still restricted to 4096 VLAN IDs that are global within its network. For many operators that are experimenting with the metro Ethernet service, the 4096 number seems good enough for an experimental network but presents a long-term roadblock if the service is to grow substantially.

Service Monitoring

Ethernet does not have an embedded mechanism that lends to service monitoring. With Frame Relay LMI, for example, service monitoring and service integrity are facilitated via messages that report the status of the PVC. Ethernet service monitoring requires additional control-plane intelligence. New Link Management Interface (LMI) protocols need to be defined and instituted between the service provider network and the CPE to allow the customer to discover the different EVCs that exist on the UNI connection. The LMI could learn the CE-VLAN to EVC map and could learn the different service parameters such as bandwidth profiles. Other protocols need to be defined to discover the integrity of the EVC in case of possible failures. You have seen in the previous section how performance parameters could indicate the availability of an EVC. Protocols to extract information from the UNI and EVC are needed to make such information usable.

Scaling the L2 Backbone

A metro carrier that is building an all-Ethernet network is at the mercy of STP. STP blocks Ethernet ports to prevent network loops. Traffic engineering (discussed in Chapter 5, "MPLS Traffic Engineering") is normally a major requirement for carriers to have control over network bandwidth and traffic trajectory. It would seem very odd for any carrier to have the traffic flow in its network be dependant on loop prevention rather than true bandwidth-optimization metrics.

Service Provisioning

Carrying a VLAN through the network is not a simple task. Any time a new carrier VLAN is created (a new VPN), care must be taken to configure that VLAN across all switches that need to participate in that VPN. The lack of any signaling protocols that allow VPN information to be exchanged makes the task manual and tedious. Early adopters of metro Ethernet have endured the pains of carrying VLANs across many switches. Even with the adoption of new protocols such as 802.1s ("Amendment to 802.1Q (TM) Virtual Bridged Local Area Networks: Multiple Spanning Trees"), the task of scaling the network is almost impossible.

Interworking with Legacy Deployments

Another challenge facing Ethernet deployments is interworking with legacy deployments such as existing Frame Relay and ATM networks. Frame Relay has been widely deployed by many enterprises as a WAN service. Remote offices are connected to headquarters via P2P Frame Relay circuits forming a hub-and-spoke topology. Enterprises that want to adopt Ethernet as an access technology expect the carrier to provide a means to connect the new sites enabled with Ethernet access with existing headquarters sites already enabled with Frame Relay. This means that a function must exist in the network that enables Frame Relay and Ethernet services to work together.

The IETF has standardized in RFC 2427, *Multiprotocol Interconnect over Frame Relay,* how to carry different protocols over Frame Relay, including Ethernet. In some other cases, the Ethernet and Frame Relay access networks are connected by an ATM core network. In this case, two service-interworking functions need to happen, one between Ethernet and ATM and another between ATM and Frame Relay. Ethernet-to-ATM interworking is achieved using RFC 2684, and ATM-to-Frame Relay interworking is achieved via the Frame Relay Forum specification FRF 8.1. Figure 3-10 illustrates the service-interworking functions.

Figure 3-10 *Service Interworking*

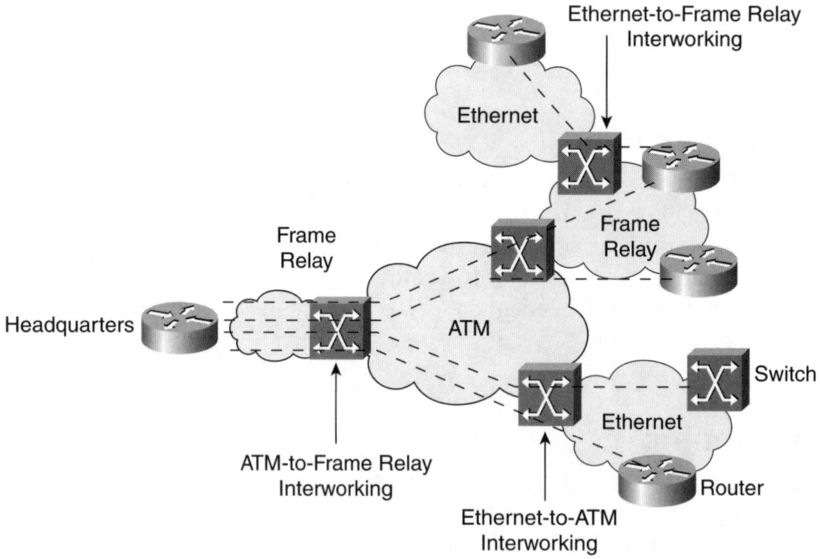

Figure 3-10 shows a scenario in which an enterprise headquarters is connected to its remote sites via Frame Relay connections carried over an ATM network. The different service-interworking functions are displayed to allow such networks to operate. For service interworking, two encapsulation methods are defined: one is bridged, and the other is routed. Both sides of the connection are either bridged or routed. Some challenges might exist if one end of the connection is connected to a LAN switch, and hence bridged, while the other end is connected to a router. Other issues will arise because of the different Address Resolution Protocol (ARP) formats between the different technologies, such as Ethernet, Frame Relay, and ATM. Some vendors are attempting to solve these problems with special software enhancements; however, such practices are still experimental and evolving.

It is all these challenges that motivated the emergence of hybrid architectures consisting of multiple L2 domains that are connected via L3 IP/MPLS cores. The network can scale because L2 Ethernet would be constrained to more-controlled access deployments that limit the VLAN and STP inefficiencies. The network can then be scaled by building a reliable IP/MPLS core. This is discussed in Chapter 4, "Hybrid L2 and L3 IP/MPLS Networks."

Conclusion

This chapter has discussed many aspects of metro Ethernet services. The MEF is active in defining the characteristics of these services, including the service definitions and framework and the many service attributes that make up the services. Defining the right traffic and performance parameters, class of service, service frame delivery, and other aspects ensures that buyers and users of the service understand what they are paying for and also helps service providers communicate their capabilities.

This chapter covers the following topics:

- Understanding VPN Components
- Delivering L3VPNs over IP
- L2 Ethernet Services over an IP/MPLS Network

CHAPTER 4

Hybrid L2 and L3 IP/MPLS Networks

In Chapter 3, "Metro Ethernet Services," you reviewed the issues that can be created by an L2-only Ethernet model. This chapter first focuses on describing a pure L3VPN implementation and its applicability to metro Ethernet. This gives you enough information to compare L3VPNs and L2VPNs relative to metro Ethernet applications. The chapter then delves into the topics of deploying L2 Ethernet services over a hybrid L2 Ethernet and L3 IP/MPLS network. Some of the basic scalability issues to be considered include restrictions on the number of customers because of the VLAN-ID limitations, scaling the L2 backbone with spanning tree, service provisioning and monitoring, and carrying VLAN information within the network. The following section describes some basic VPN definitions and terminology.

Understanding VPN Components

There are normally two types of VPNs: customer premises equipment-(CPE) based VPNs and network-based VPNs. With CPE-based VPNs, secure connections are created between the different customer premises equipment to form a closed user group/VPN. This normally creates scalability issues, because many CPE devices have to be interconnected in a full mesh or a partial mesh to allow point-to-multipoint connectivity. On the other hand, network-based VPNs create some level of hierarchy where connections from many CEs are aggregated into an edge switch or router offering the VPN service.

The definitions of the different elements of the network follow:

- **Customer edge (CE)**—The customer edge device resides at the edge of the enterprise. This device is usually a router or a host in L3VPNs; however, as you will see with L2VPNs, the CE could also be an L2 switch. The CE connects to the provider network via different data-link protocols such as PPP, ATM, Frame Relay, Ethernet, GRE, and so on.
- **Provider edge (PE)**—The provider edge device is a provider-owned device that offers the first level of aggregation for the different CEs. The PE logically separates the different VPNs it participates in. The PE does not have to participate in all VPNs but would only participate in the VPNs of the enterprises that are directly attached to it.

- **Provider (P)**—The provider device is normally a core IP/MPLS router that offers a second level of aggregation for the PEs. This device does not participate in any VPN functionality and is normally agnostic to the presence of any VPNs.

The remainder of the chapter mainly focuses on different types of VPNs and how they differ between an L2 or L3 service. The different types of VPNs include

- GRE- and MPLS-based L3VPNs
- Hybrid Ethernet and IP/MPLS L2VPNs via L2TPv3, Ethernet over MPLS (EoMPLS), and Virtual Private LAN Service (VPLS)

Delivering L3VPNs over IP

L3VPNs allow the provider to extend its customer's private IP network over the provider's backbone. When delivering an L3 service, the service provider is normally involved in the assignment and management of a pool of IP addresses allocated to its customer. This is typical of carriers that are also ISPs offering Internet services or carriers offering IP multicast services and so on. L3VPNs can be delivered via GRE tunnels or MPLS L3VPNs.

GRE-Based VPNs

L3VPN services over IP have traditionally been done using generic routing encapsulation (GRE) tunnels, which allow the encapsulation of IP packets inside IP packets. GRE-based VPNs are CE-based VPNs. A network hierarchy can be maintained in which an enterprise that has, for example, a private IP addressing scheme can create a private VPN on top of a service provider's network. IP forwarding is used to exchange traffic between the endpoints of GRE tunnels, allowing full or partial connectivity between the different sites of the same enterprise. From a scalability perspective, this scheme could scale to a certain point and then become unmanageable, because the VPN becomes the collection of many point-to-point tunnels. As many sites are added to the VPN and many tunnels have to be created to all or a partial set of the other sites, the operational management of such a scheme becomes cost-prohibitive, especially because there are no rules or guidelines or an industry push to allow such tunneling schemes to scale.

Figure 4-1 shows an example of a service provider delivering a GRE-based VPN service using managed CEs located at different enterprise sites. The provider is managing the CEs at each site of each enterprise and is managing the tunnel connectivity between the different sites. As the number of enterprises grows and the number of sites per enterprise grows as well, this model will definitely have scalability issues. Notice that different enterprises could use overlapping private IP addresses, because all IP and routing information between the enterprise sites is carried within tunnels and hence is hidden from the provider's network and other enterprises.

For large-scale deployments of IP VPNs, the industry has gradually moved toward adopting MPLS L3VPNs, as defined in RFC 2547.

Figure 4-1 *GRE Tunnels*

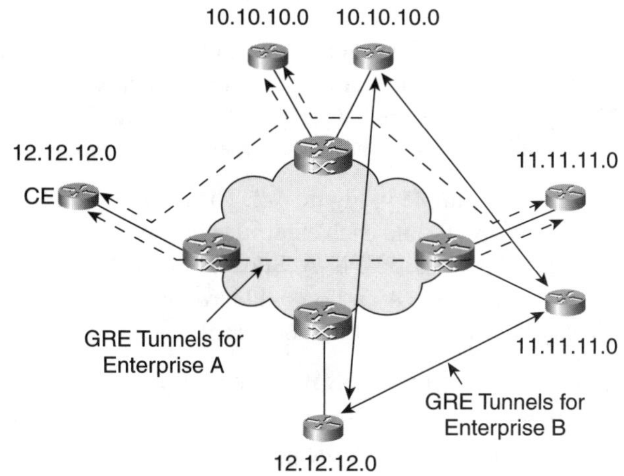

MPLS L3VPNs

MPLS L3VPNs are network-based VPNs. This scheme defines a scalable way for the service provider to offer VPN services for enterprises. Enterprises can leverage the service provider backbone to globally expand their intranets and extranets. An *intranet* normally means that all sites in the VPN connect to the same customer, and *extranet* means that the various sites in the VPN are owned by different enterprises, such as the suppliers of an enterprise. An example of an extranet would be a car manufacturer that builds a network that connects it and all its parts suppliers in a private network.

Although MPLS L3VPNs provide a sound and scalable solution for delivering VPNs over IP, they have some characteristics that make them overkill for metro Ethernet services. L3VPNs, for example, are more adequate for delivering IP services than L2VPN services. This is one of the reasons that the industry is looking at L2VPNs for metro Ethernet services. To understand the differences between L2VPNs and L3VPNs, it helps to identify the different elements of MPLS L3VPNs (RFC 2547) and the challenges that come with them.

MPLS L3VPNs use the CE, PE, and P terminology described earlier in the "Understanding VPN Components" section. In the case where the CE is a router, the CE and PE become routing peers if a routing protocol is used between the two to exchange IP prefixes. In other scenarios, static routing is used between the PE and CE to alleviate the exchange of routing information. With L3VPNs, enterprise edge routers have to talk only to their direct neighbor, which is the router owned by the provider. From a scalability perspective, the L3VPN model scales very well, because each site does not need to know of the existence of other sites. On the other hand, this model is not so good for enterprises that would like to maintain their own internal routing practices and control the routing mechanism used between the different sites. Also, this model forces the service provider to participate in and manage the IP addressing schemes for its customers, as is typically done when IP services are sold. This model is not adequate for selling L2 services only (L2VPN) where the customer's IP network becomes an overlay on top of the service provider's network.

Another disadvantage of L3VPNs when used for metro Ethernet services is that L3VPNs apply only to the transport of IPv4 packets. For metro deployments, enterprise traffic consists of IPv4 as well as other types of traffic such as IPX and SNA. An L2VPN allows any type of traffic to be encapsulated and transported across the metro network.

Take a close look at the example in Figure 4-2. The provider is delivering VPN services to two different enterprises, A and B, and each enterprise has two different sites. Sites A1 and A2 are part of enterprise A and belong to VPN-A. Sites B1 and B2 are part of enterprise B and belong to VPN-B. Note that enterprises A and B could have overlapping IP addresses. The following are the reasons why the MPLS L3VPN model scales:

- Each PE knows only of the VPNs it attaches to. PE1 knows only of VPN-A, and PE3 knows only of VPN-B.
- The P routers do not have any VPN information.
- The CE routers peer with their directly attached PEs. A1 peers with PE1, B1 peers with PE3, and so on.

Figure 4-2 *MPLS L3VPN Principles*

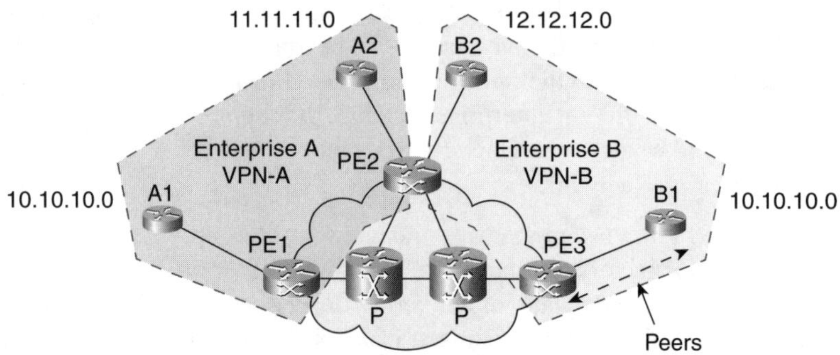

The following sections describe

- How MPLS L3VPN PEs maintain separate forwarding tables between different VPNs
- The concept of VPN-IPv4 addresses
- How packets are transported across the backbone using the MPLS L3VPN mechanism

Maintaining Site Virtual Router Forwarding Tables

The fundamental operation of the MPLS L3VPN model follows:

- Each PE router maintains a separate virtual router forwarding (VRF) table for each site the PE is attached to. The forwarding table contains the routes to all other sites that participate in a set of VPNs.

- The PEs populate the forwarding tables from information learned from the directly attached sites or learned across the backbone from other PEs that have a VPN in common. Information from directly attached CEs is learned via routing protocols such as OSPF, IS-IS, RIP, and BGP or via static configuration. Distribution of VPN information across the backbone is done via *multiprotocol BGP (MP-BGP).* MP-BGP introduces extensions to the BGP-4 protocol to allow IPv4 prefixes that are learned from different VPNs to be exchanged across the backbone. IP prefixes can overlap between different VPNs via the use of VPN-IPv4 address, as explained later, in the section "Using VPN-IPv4 Addresses in MPLS L3VPNs."
- The CEs learn from the PEs about the routes they can reach via routing protocols or static configuration.

Traffic is forwarded across the backbone using MPLS. MPLS is used because the backbone P routers have no VPN routes; hence, traditional IP routing cannot be used. Figure 4-3 illustrates the packet forwarding process.

Figure 4-3 *The Packet Forwarding Process*

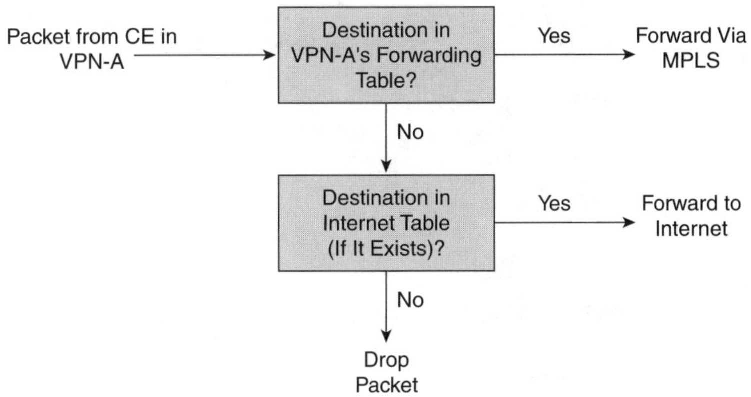

As Figure 4-3 shows, when a packet is received from a site, the PE looks up the IP destination in the site's forwarding table and, if found, forwards the packet via MPLS. Otherwise, the PE checks the destination in other forwarding tables and discards the packet if no match has been made.

Figure 4-4 shows how the PEs maintain a different forwarding table per site. PE1 contains a forwarding table for enterprise A site 1 (A1). That forwarding table is populated from routes learned from A1 and from BGP routes across the backbone. PE2 contains a forwarding table for both enterprise A site 2 (A2) and enterprise B site 2 (B2).

MPLS L3VPNs use target VPNs, VPN of origin, and site of origin to be able to identify and separate the different routes belonging to different VPNs and to clearly identify the origin of a particular route. The following sections describe these features.

Figure 4-4 *PE Logical Separation*

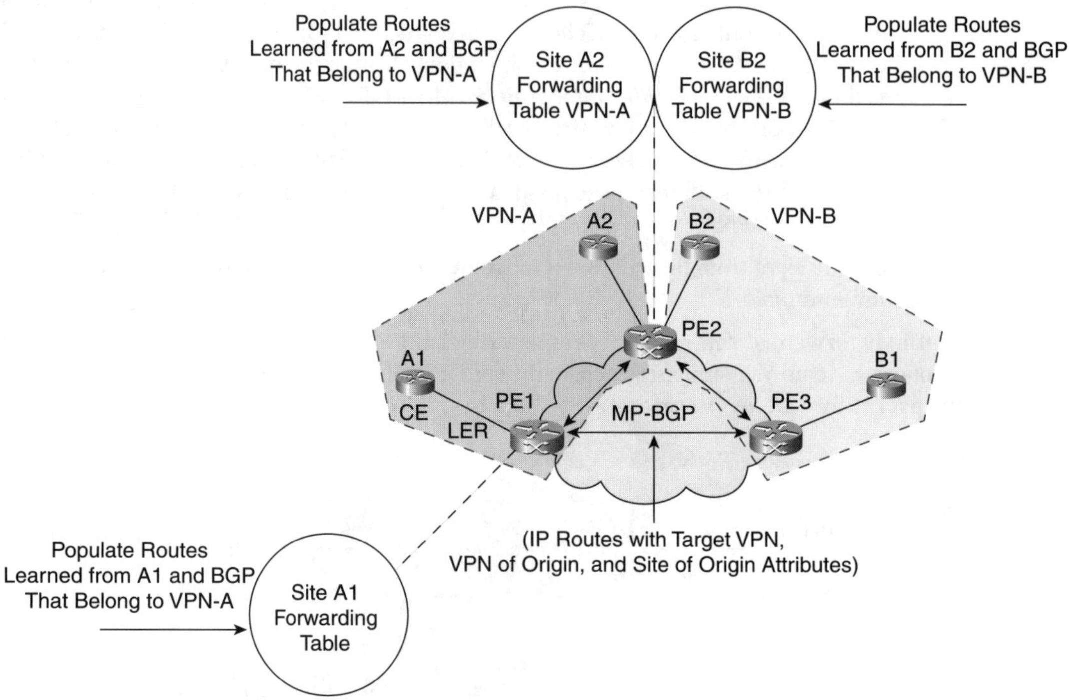

Target VPN

For identifying different VPNs, every per-site forwarding table is associated with one or more target VPN attributes. When a PE router creates a VPN-IPv4 route, the route is associated with one or more target VPN attributes, which are carried in BGP as attributes of the route. A *route attribute* is a parameter that gives the route special characteristics and is a field that is distributed inside a BGP advertisement. The target VPN attribute identifies a set of sites. Associating a particular target VPN attribute with a route allows the route to be placed in the per-site forwarding tables used for routing traffic that is received from the corresponding site. In Figure 4-4, when PE1 receives BGP routes from PE2, PE1 installs in the A1 forwarding table only routes that have a target VPN attribute VPN-A. This ensures that PE1 does not contain any routes to VPN-B, because PE1 does not have any attached sites that belong to VPN-B. On the other hand, PE2 installs in the respective forwarding tables routes that belong to VPN-A and VPN-B, because PE2 is attached to sites that belong to both VPNs.

NOTE In the context of MPLS L3VPN, IPv4 addresses are referred to as VPN-IPv4 addresses. The section "Using VPN-IPv4 Addresses in MPLS L3VPNs" discusses scenarios for VPN-IPv4 addresses in more detail.

VPN of Origin

Additionally, a VPN-IPv4 route may be optionally associated with a VPN of origin attribute. This attribute uniquely identifies a set of sites and identifies the corresponding route as having come from one of the sites in that set. Typical uses of this attribute might be to identify the enterprise that owns the site where the route leads, or to identify the site's intranet. However, other uses are also possible, such as to identify which routes to accept and which to drop based on the VPN of origin. By using both the target VPN and the VPN of origin attributes, different kinds of VPNs can be constructed.

Site of Origin

Another attribute, called the site of origin attribute, uniquely identifies the site from which the PE router learned the route (this attribute could be encoded as an extended BGP community attribute). All routes learned from a particular site must be assigned the same site of origin attribute.

Using VPN-IPv4 Addresses in MPLS L3VPNs

The purpose of VPN-IPv4 addresses is to allow routers to create different routes to a common IPv4 address. This is useful in different scenarios that relate to L3VPNs.

One such scenario occurs when multiple VPNs have overlapping IPv4 addresses. In this case, routers need to treat each address differently when populating the per-site forwarding table. If the same address belongs in two different VPNs, the router needs to place the same address into two different VRF tables. Another use of VPN-IPv4 is to create separate routes to reach the same IPv4 destination address. In the case of an enterprise that has an intranet and an extranet, the same server can have its IP address advertised in two different routes, one used by the intranet and another by the extranet. The extranet route could be forced to go through a firewall before reaching the server.

The VPN-IPv4 address, as shown in Figure 4-5, is a 12-byte quantity, beginning with an 8-byte Route Distinguisher (RD) and ending with a 4-byte IPv4 address. The RD consists of a Type field that indicates the length of the Administrator and Assigned Number fields. The Administrator field identifies an Assigned Number authority field, such as an autonomous system number given to a certain service provider. The service provider can then allocate the assigned number to be used for a particular purpose. Note that the RD by itself does not contain enough information to indicate the origin of the route or to which VPNs the route needs to be distributed. In other words, the RD is not indicative of a particular VPN. The purpose of the RD is only to allow the router to create different routes to the same IPv4 address.

As referenced in Figure 4-5, ISP A wants to distinguish between two IPv4 addresses 10.10.10.0; therefore, it assigns these two addresses two different RDs. The RD administrator number is ISP A's autonomous system (AS) number (1111). The assigned numbers 1 and 2 are just arbitrary numbers that help the routers distinguish between the two IP addresses that could be in the same VPN or different VPNs. Again, it is important to understand that the VPN-IPv4 does not modify the IP address itself but rather is an attribute sent within BGP (RFC 2283) that indicates that the IP address belongs to a certain family.

Figure 4-5 *VPN-IPv4 Address*

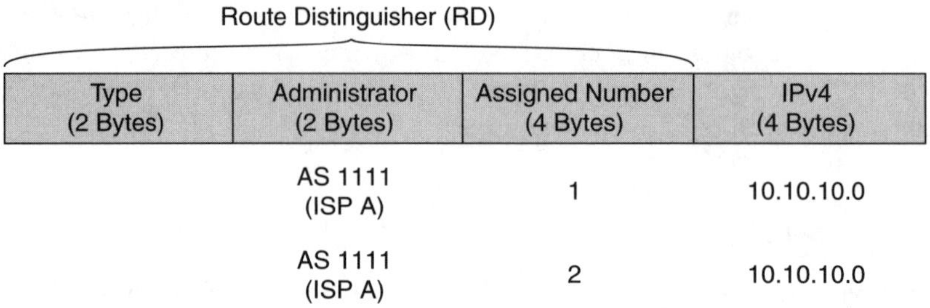

Forwarding Traffic Across the Backbone

Only the edge PE routers have information about the VPN IP prefixes. The backbone P routers do not carry any VPN IP prefixes. With traditional IP forwarding, this model does not work, because the P routers drop any traffic destined for the VPN IP addresses. MPLS is used to allow packet forwarding based on labels rather than IP addresses. The PE routers tag the traffic with the right label based on the destination IP address it needs to go to, and the MPLS P routers switch the traffic based on the MPLS labels. If this model is not adopted, the P routers would have to carry IP prefixes for all VPNs, which would not scale.

MPLS L3VPN does not mandate the use of traffic engineering (a topic that is explained in more detail in Chapter 5, "MPLS Traffic Engineering"). When traffic moves from one site to another across the carrier's backbone, it follows the MPLS label switched path (LSP) assigned for that traffic. The LSP itself could have been formed via dynamic routing calculated by the routing protocols. On the other hand, the LSP could be traffic-engineered to allow certain types of traffic to follow a well-defined trajectory. Also, many mechanisms can be used for traffic rerouting in case of failure. The mechanism used depends on whether the carrier requires normal IP routing or MPLS fast reroute mechanisms (as explained in Chapter 6, "RSVP for Traffic Engineering and Fast Reroute").

The traffic across the MPLS backbone carries a label stack. The label on top of the stack is called the packet-switched network (PSN) tunnel label and is indicative of the path that a packet needs to take from the ingress PE to the egress PE. The label beneath is indicative of the particular VPN that the packet belongs to. In the case where IP forwarding (rather than MPLS) is used in the provider routers, the PSN tunnel can be replaced by a GRE tunnel, and the packet would carry the VPN label inside the GRE tunnel.

Applicability of MPLS L3VPNs for Metro Ethernet

The MPLS L3VPN model presents many challenges if used to deliver metro Ethernet services. This model is more applicable for delivering IP services, where an enterprise is outsourcing the operation of its WAN/metro IP network to a service provider. From an administration point of view, the MPLS L3VPN model dictates that the carrier is involved with the customer's IP addressing scheme. Remember that the CE routers would have to peer with the provider's routers. If static routing is not used, routing exchange between the PE and CE might involve configuring routing protocols like RIP and OSPF and will involve many guidelines to allow protocols such as OSPF to understand the separation between the different VPN routes and to distribute the correct routes between BGP and OSPF.

From an equipment vendor perspective, while the MPLS L3VPN model scales in theory, it introduces major overhead on the edge routers. If you assume that an edge router needs to support 1000 VPNs, and each VPN has 1000 IP prefixes, the edge routers would have to maintain at least 1,000,000 IP prefixes in 1000 separate forwarding tables. Most routers on the market are still struggling to reach 256,000 to 500,000 IP entries, depending on the vendor's implementation. So what would happen if the IP prefixes per VPN reaches 5000 entries rather than 1000? A clever answer would be to support 200 VPNs per PE router to stay within the 1,000,000 prefix limit, until vendors find a way to increase that number.

There are other ways of deploying L3VPNs, such as using virtual routers where different instances of routing protocols run on each router. Each routing instance carries the IP prefixes of a different VPN, and traffic is forwarded across the network using traditional IP forwarding; hence, the final outcome is very similar to running MPLS L3VPNs. The MPLS L3VPN and virtual routers have their advantages in delivering IP services, which include IP QoS mechanisms, IP address pool management, and so on. These advantages are very important but are outside the scope of this book and will not be discussed. However, L3VPN is still overkill for deploying metro Ethernet services, which focuses on simpler deployments and L2 services.

It is understandable why the industry started looking at simpler VPN schemes like L2VPNs to avoid many of the L3VPN complexities and to create a model in which simpler services like Transparent LAN Service (TLS) can be deployed with less operational overhead.

In an L2 service, the carrier offers its customers the ability to "transparently" overlay their own networks on top of the carrier's network. The customer of a carrier could be an ISP that offers Internet services and purchases last-mile connectivity from the carrier, or an enterprise customer that uses the carrier's backbone to build the enterprise WAN while still controlling its internal IP routing.

L2 Ethernet Services over an IP/MPLS Network

The inherent properties of an IP/MPLS network mitigate most of the scalability issues by design. IP and MPLS have been widely deployed in large service provider networks, and these protocols have been fine-tuned over the years to offer high levels of stability and flexibility. Table 4-1 shows a brief comparison of the merits of IP/MPLS versus L2 Ethernet networks.

Table 4-1 *Comparing Ethernet and IP/MPLS*

Feature	Ethernet	IP/MPLS
Signaling	No signaling	LDP, RSVP-TE, and so on
Loop-free topology	Blocked ports via Spanning Tree Protocol	Yes, via routing protocols and Time To Live
User and service identification	VLAN ID space limited	Label space more scalable
Traffic engineering (TE)	No TE	RSVP-TE
Restoration	Via STP	Backup path, MPLS fast reroute
Address aggregation	No aggregation for MAC addresses	Yes, via classless interdomain routing

Segmenting the L2 Ethernet network with IP/MPLS creates an L2 Ethernet domain at the metro access and an IP/MPLS metro edge/core and WAN backbone capable of carrying the L2 services. As you will see in this chapter, the closer the IP/MPLS network gets to the customer, the more scalable the service becomes; however, it introduces more complications.

The exercise of deploying Ethernet L2 services becomes one of balance between the L2 Ethernet simplicity and its scalability shortfalls and the L3 IP/MPLS scalability and its complexity shortfalls. First, it helps to compare and contrast the benefits that IP/MPLS offers over flat L2 networks.

You have seen so far in this book two extremes: one with an MPLS L3VPN service and one with an all-L2 Ethernet service. In this chapter, you see the hybrid model that falls in between. Figure 4-6 shows how an IP/MPLS domain can create a level of hierarchy that allows the L2 services to be confined to the access/edge network. There could be either an L2 access with IP/MPLS edge and core or an L2 access and edge with IP/MPLS core.

Figure 4-6 *Hybrid L2 and IP/MPLS Metro*

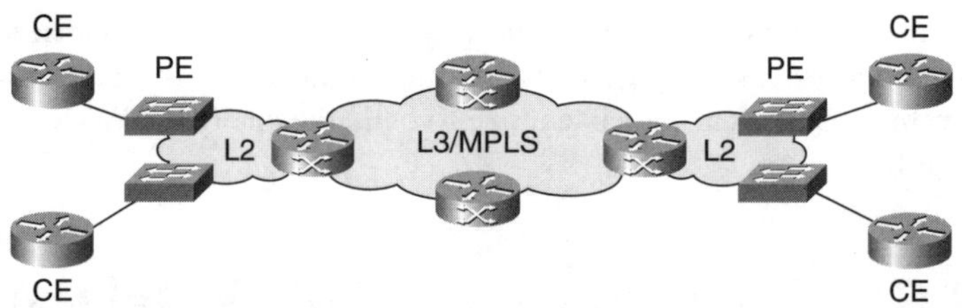

The IP/MPLS edge/core network limits the L2 domains to the access or access/edge side and provides a scalable vehicle to carry the L2 services across.

The L2 Ethernet service across an IP/MPLS cloud can be a point-to-point (P2P) or multipoint-to-multipoint (MP2MP) service. This is very similar to the Metro Ethernet Forum (MEF) definitions of an Ethernet Line Service (ELS) and Ethernet LAN Service (E-LAN). The following associates the service with the different methods to deliver it:

- **P2P Ethernet Service**—Comparable to ELS, delivered via:
  - L2TPv3 over an IP network
  - Ethernet over MPLS, also known as draft-martini in reference to the author of the original draft
- **MP2MP Ethernet Service**—Comparable to E-LAN, delivered via VPLS

Before getting into more details of the different mechanisms to deploy P2P and MP2MP L2 services, it helps to understand the packet leased-line concept, which is also referred to as pseudowire (PW), as explained next.

The Pseudowire Concept

The Internet Engineering Task Force (IETF) has defined the concept of a pseudowire. An Ethernet PW allows Ethernet/802.3 protocol data units (PDUs) to be carried over a PSN, such as an IP/MPLS network. This allows service providers or enterprise networks to leverage an existing IP/MPLS network to offer Ethernet services.

You could set up the PW via manual configuration or a signaling protocol such as BGP or LDP. The PW may operate over an MPLS, IPv4, or IPv6 PSN.

An Ethernet PW emulates a single Ethernet link between exactly two endpoints. The PW terminates a logical port within the PE. This port provides an Ethernet MAC service that delivers each Ethernet packet that is received at the logical port to the logical port in the corresponding PE at the other end of the PW. Before a packet is inserted into the PW at the PE, the packet can go through packet processing functions that may include the following:

- Stripping
- Tag stacking or swapping
- Bridging
- L2 encapsulation
- Policing
- Shaping

Figure 4-7 shows a reference model that the IETF has adopted to support the Ethernet PW emulated services. As Figure 4-7 shows, multiple PWs can be carried across the network inside a bigger tunnel called the PSN tunnel. The PSN tunnel is a measure to aggregate multiple PWs into a single tunnel across the network. The PSN tunnel could be formed using generic routing encapsulation (GRE), Layer 2 Tunneling Protocol (L2TP), or MPLS and is a way to shield the internals of the network, such as the P routers, from information relating to the service provided by the PEs. In Figure 4-7, while the PE routers are involved in creating the PWs and mapping the L2 service to the PW, the P routers are agnostic to the L2 service and are passing either IP or MPLS packets from one edge of the backbone to the other.

Figure 4-7 *Creating Pseudowires*

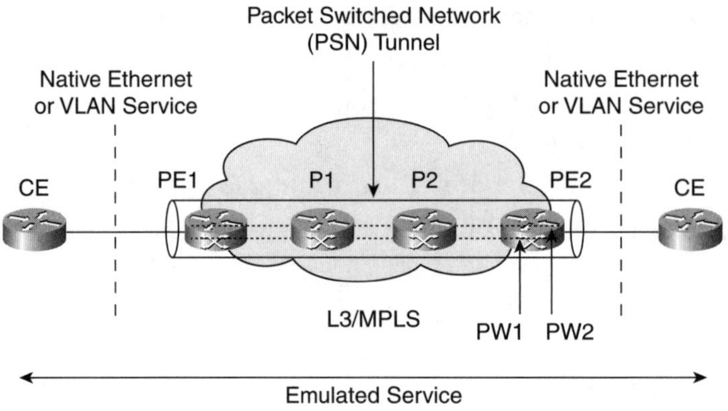

The following sections describe the different mechanisms used to deliver P2P and MP2MP L2 Ethernet service over MPLS, starting with L2TPv3. You then learn more about Ethernet over MPLS—draft-martini and VPLS.

PW Setup Via L2TPv3

L2TP provides a dynamic tunneling mechanism for multiple L2 circuits across a packet-oriented data network. L2TP was originally defined as a standard method for tunneling the Point-to-Point Protocol (PPP) and has evolved as a mechanism to tunnel a number of other L2 protocols, including Ethernet. L2TP as defined in RFC 2661, *Layer 2 Tunneling Protocol (L2TP),* is referred to as L2TPv2. L2TPv3 is an extension of that protocol that allows more flexibility in carrying L2 protocols other than PPP. Notable differences between L2TPv2 and L2TPv3 are the separation of all PPP-related attributes and references and the transition from a 16-bit Session ID and Tunnel ID to a 32-bit Session ID and Control Connection ID, offering more scalability in deploying L2 tunnels.

With L2TPv3 as the tunneling protocol, Ethernet PWs are actually L2TPv3 sessions. An L2TP control connection has to be set up first between two L2TP control connection endpoints (LCCEs) at each end, and then individual PWs can be established as L2TP sessions.

The provisioning of an Ethernet port or Ethernet VLAN and its association with a PW on the PE triggers the establishment of an L2TP session. The following are the elements needed for the PW establishment:

- **PW type**—The type of PW can be either Ethernet port or Ethernet VLAN. The Ethernet port type allows the connection of two physical Ethernet ports, and the Ethernet VLAN indicates that an Ethernet VLAN is connected to another Ethernet VLAN.
- **PW ID**—Each PW is associated with a PW ID that identifies the actual PW.

The entire Ethernet frame without the preamble or FCS is encapsulated in L2TPv3 and is sent as a single packet by the ingress side of the L2TPv3 tunnel. This is done regardless of whether an 802.1Q tag is present in the Ethernet frame. For a PW of type Ethernet port, the egress side of the tunnel simply de-encapsulates the Ethernet frame and sends it out on the appropriate interface without modifying the Ethernet header. The Ethernet PW over L2TP is homogeneous with respect to packet encapsulation, meaning that both ends of the PW are either VLAN tagged or untagged; however, once the packet leaves the PW, a Native Service Processing (NSP) function within the PE can still manipulate the tag information. For VLAN-to-VLAN connectivity, for example, the egress NSP function may rewrite the VLAN tag if a tag replacement or swapping function is needed.

NOTE The preamble is a pattern of 0s and 1s that tells a station that an Ethernet frame is coming. FCS is the frame check sequence that checks for damage that might have occurred to the frame in transit. These fields are not carried inside the PW.

Figure 4-8 shows an L2TP control connection formed between PE1 and PE2. Over that connection two PWs or L2TPV3 sessions are formed. The two sessions are of type Ethernet VLAN, which means that the PW represents a connection between two VLANs. For session 1, VLAN 10 has been left intact on both sides. For session 2, the NSP function within PE2 rewrites VLAN ID 20 to VLAN ID 30 before delivering the packet on the local segment.

Figure 4-8 *Ethernet over L2TPV3*

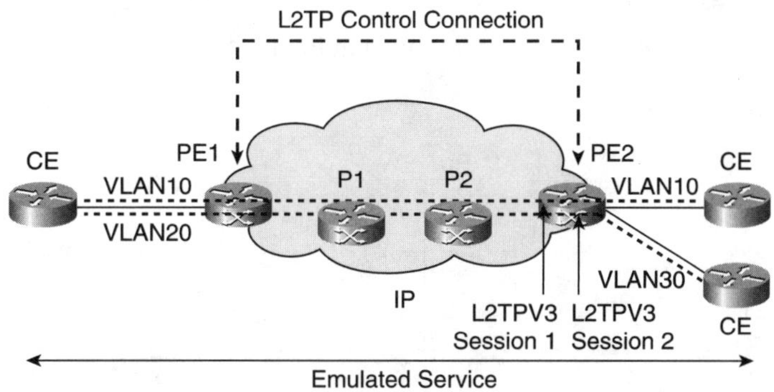

Ethernet over MPLS—Draft-Martini

You have seen in the previous section how an Ethernet packet can be transported using an L2TPv3 tunnel over an IP network. The IETF has also defined a way to carry L2 traffic over an MPLS network. This includes carrying Ethernet over MPLS (EoMPLS), Frame Relay, and ATM. This is also referred to as "draft-martini" encapsulation in reference to the author of the original Internet draft that defined Layer 2 encapsulation over MPLS. With this type of encapsulation, PWs are constructed by building a pair of unidirectional MPLS virtual connection (VC) LSPs between the two PE endpoints. One VC-LSP is for outgoing traffic, and the other is for incoming traffic. The VC-LSPs are identified using MPLS labels that are statically assigned or assigned using the Label Distribution Protocol (LDP).

EoMPLS uses "targeted" LDP, which allows the LDP session to be established between the ingress and egress PEs irrespective of whether the PEs are adjacent (directly connected) or nonadjacent (not directly connected). The following section explains the mechanism of encapsulating the Ethernet frames over the MPLS network and shows two scenarios of using LDP to establish PWs between directly connected and non-directly connected PEs.

Ethernet Encapsulation

Ethernet encapsulation is very similar to what was described in the "PW Setup Via L2TPv3" section, but a different terminology is introduced. The entire Ethernet frame without any preamble or FCS is transported as a single packet over the PW. The PW could be configured as one of the following:

- **Raw mode**—In raw mode, the assumption is that the PW represents a virtual connection between two Ethernet ports. What goes in one side goes out the other side. The traffic could be tagged or untagged and comes out on the egress untouched.
- **Tagged mode**—In tagged mode, the assumption is that the PW represents a connection between two VLANs. Each VLAN is represented by a different PW and is switched differently in the network. The tag value that comes in on ingress might be overwritten on the egress side of the PW.

The raw and tagged modes are represented in Figure 4-9.

Figure 4-9 *Martini Tunnel Modes*

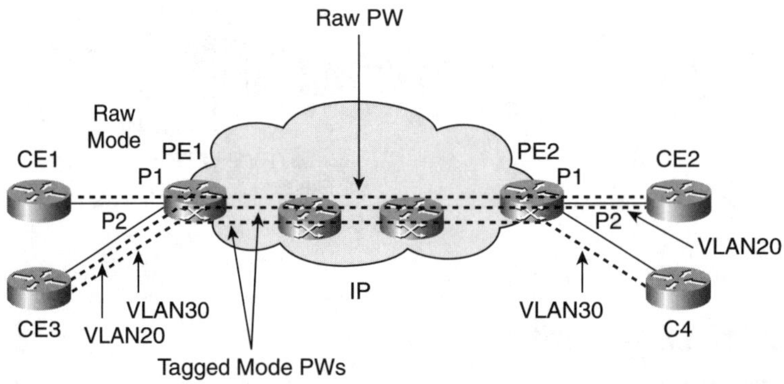

Figure 4-9 shows that PE1 has established a PW of type raw with PE2 over which all traffic coming in on port 1 is mapped. As such, the traffic comes out as-is at the other end of the PW on PE2 port 1. Also, PE1 has defined on port 2 two PWs of type tagged. The first PW maps VLAN 20 on PE1 port 2 and connects it to VLAN 20 on PE2 port 1, and the second PW maps VLAN 30 on PE1 port 2 and maps it to VLAN 30 on PE2 port 2.

Maximum Transmit Unit

Both ends of the PW must agree on their maximum transmission unit size to be transported over the PSN, and the network must be configured to transport the largest encapsulation frames. If MPLS is used as the tunneling protocol, the addition of the MPLS shim layer increases the frame size. If the vendor implementation does not support fragmentation when tunneling the Ethernet service over MPLS, care must be taken to ensure that the IP/MPLS routers in the network are adjusted to the largest maximum transmission unit.

Frame Reordering

The IEEE 802.3 requires that frames from the same conversation be delivered in sequence. Because the frames are now encapsulated inside PWs, the PW must be able to support frame reordering.

Using LDP with Directly Connected PEs

Figure 4-10 shows how martini tunnels can be established using LDP between two directly connected PEs, such as PE1 and PE2. First, an LDP session needs to be established between the two PEs. Once the LDP session has been established, all PWs are signaled over that session. In this example, you can see the establishment of one bidirectional PW via two unidirectional VC-LSPs. Once both VC-LSPs are established, the PW is considered operational. PE2 assigns label 102 and sends it to PE1 to be used for propagating traffic from PE1 to PE2. In turn, PE1 assigns label 201 and sends it to PE2 to be used for propagating traffic from PE2 to PE1. The label is pushed into the data packet before transmission, and it indicates to the opposite endpoint what to expect regarding the encapsulated traffic. Remember that this type of encapsulation is used to tunnel not only Ethernet but other types of traffic such as ATM, Frame Relay, and Circuit Emulation traffic. The VC label gives the opposite side an indication of how to process the data traffic that is coming over the VC-LSP.

Figure 4-10 *LDP Between Directly Connected PEs*

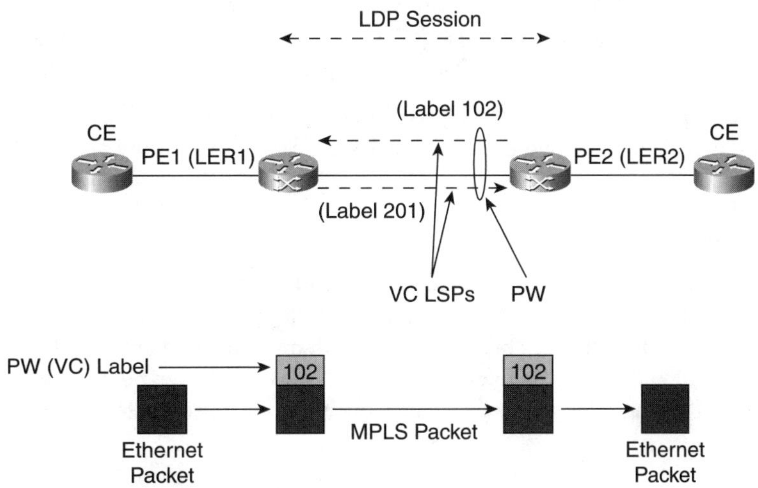

The VC information is carried in a label mapping message sent in downstream unsolicited mode with a new type of forwarding equivalency class element defined as follows (refer to Figure 4-11):

- **VC Type**—A value that represents whether the VC is of type Frame Relay data-link connection identifier (DLCI), PPP, Ethernet tagged or untagged frames, ATM cell, Circuit Emulation, and so on.
- **PW ID or VC ID**—A connection ID that together with the PW type identifies a particular PW (VC). For P2P tunnels, the VC ID gives an indication of a particular service. You will see in the next section that in the context of an MP2MP VPLS service, the VC ID is indicative of a LAN.
- **Group ID**—Represents a group of PWs. The Group ID is intended to be used as a port index or a virtual tunnel index. The Group ID can simplify configuration by creating a group membership for all PWs that belong to the same group, such as an Ethernet port carrying multiple PWs.
- **Interface Parameters**—A field that is used to provide interface-specific parameters, such as the interface maximum transmission unit.

Figure 4-11 *LDP Forwarding Equivalency Class*

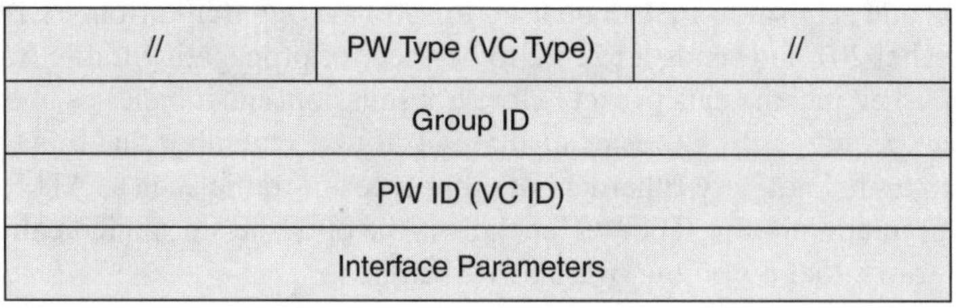

MPLS PWs are formed using two unidirectional VC-LSPs, which means that for each PW that is established from ingress to egress, a "matching" PW needs to be established between egress and ingress with the same PW ID and PW type.

In the remainder of this book, the terms PW and VC-LSP are used interchangeably, but remember that a PW is formed of two unidirectional VC-LSPs, one inbound and one outbound.

Non-Directly Connected PEs

If the PEs are not directly connected, the PE-to-PE traffic has to traverse the MPLS backbone across P core routers. These routers do not need to get involved with the different services offered at the edge and are concerned only with transporting the traffic from PE to PE. To hide the information from the P routers, LSP tunnels are constructed between the different PEs using targeted LDP, and the different PWs can share these tunnels. The construction of the LSP tunnels does not relate to the Ethernet MPLS service whatsoever. These tunnels can be constructed via different methods, such as GRE, L2TP, or MPLS. If constructed via MPLS, a signaling protocol such as RSVP-TE can be constructed to traffic-engineer these LSP tunnels across the network (RSVP-TE is explained in Chapter 6.

In Figure 4-12, an LSP tunnel, called a packet-switched network (PSN) tunnel LSP, is constructed between PE1 and PE2, and the PW is carried across that tunnel. The PSN tunnel LSP is constructed by having PE1 push a tunnel label that gets the packets from PE1 to PE2. The PSN tunnel label is pushed on top of the VC label, which gives the other side an indication of how to process the traffic. The P routers do not see the VC label and are only concerned with switching the traffic between the PE routers irrespective of the service (indicated by the VC labels) that is carried. The following describes the process of transporting a packet from ingress PE to egress PE:

1 When PE1 sends a Layer 2 PDU to PE2, it first pushes a VC label on its label stack and then a PSN tunnel label.

2 As shown in Figure 4-12, a targeted LDP session is formed between PE1 and PE2.

Figure 4-12 *LDP Between Non-Directly Connected PEs*

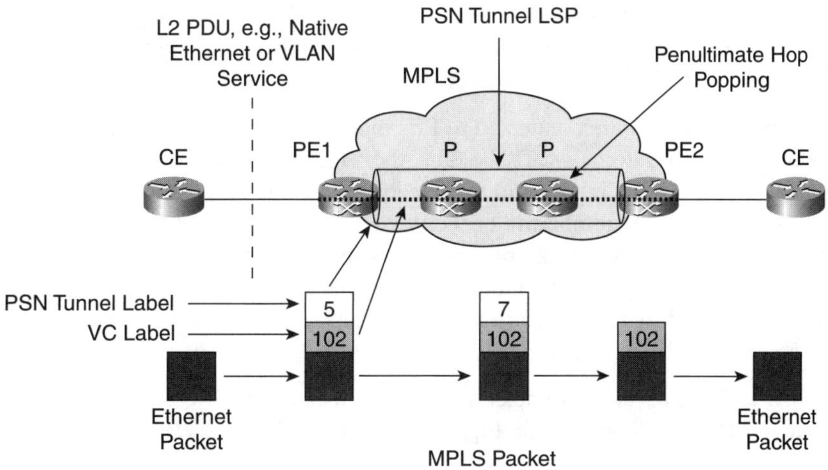

3 PE2 gives PE1 label 102 to be used for traffic going from PE1 to PE2 (the same scenario happens in the reverse direction).

4 Label 102 is pushed by PE1, and then a PSN tunnel LSP label 5 is pushed on top.

5 The P routers use the upper label to switch the traffic toward PE2. The P routers do not have visibility to the VC labels.

6 The last router before PE2 performs a penultimate hop popping function to remove the upper label before it reaches PE2. *Penultimate hop popping* is a standard MPLS function that alleviates the router at the end of the LSP (PE2 in this case) from performing a popping function and examining the traffic beneath at the same time. PE2 receives the traffic with the inner label 102, which gives an indication of what is expected in the PW.

So far you have seen a P2P L2 service over MPLS. Next, MP2MP is discussed when a LAN is emulated over MPLS using VPLS.

Virtual Private LAN Service

With Virtual Private LAN Service, an L2VPN emulates a LAN that provides full learning and switching capabilities. Learning and switching are done by allowing PE routers to forward Ethernet frames based on learning the MAC addresses of end stations that belong to the VPLS. VPLS allows an enterprise customer to be in full control of its WAN routing policies by running the routing service transparently over a private or public IP/MPLS backbone. VPLS services are transparent to higher-layer protocols and use L2 emulated LANs to transport any type of traffic, such as IPv4, IPv6, MPLS, IPX, and so on.

VPLS is flexible because it emulates a LAN, but by doing so it has all the limitations of Ethernet protocols, including MAC addresses, learning, broadcasts, flooding, and so on. The difference between VPLS and EoMPLS is that VPLS offers an MP2MP model instead of the previously discussed P2P model with L2TPv3 or EoMPLS using martini tunnels.

With VPLS, the CEs are connected to PEs that are VPLS-capable. The PEs can participate in one or many VPLS domains. To the CEs, the VPLS domains look like an Ethernet switch, and the CEs can exchange information with each other as if they were connected via a LAN. This also facilitates the IP numbering of the WAN links, because the VPLS could be formed with a single IP subnet. Separate L2 broadcast domains are maintained on a per-VPLS basis by PEs. Such domains are then mapped into tunnels in the service provider network. These tunnels can either be specific to a VPLS (for example, IP tunnels) or shared among several VPLSs (for example, with MPLS LSPs).

The PE-to-PE links carry tunneled Ethernet frames using different technologies such as GRE, IPSec, L2TP, MPLS, and so on. Figure 4-13 shows a typical VPLS reference model.

As Figure 4-13 shows, MPLS LSP tunnels are created between different PEs. These MPLS tunnels can be shared among different VPLS domains and with other services such as EoMPLS tunnels, Layer 3 MPLS VPN tunnels, and so on. The PE routers are configured to be part of one, many, or no VPLS, depending on whether they are participating in a VPLS service.

NOTE The access network connecting the CEs to the PEs could be built with Ethernet technology or with next-generation SONET/SDH running Ethernet framing over the Generic Framing Protocol (GFP) or any logical links such as ATM PVCs or T1/E1 TDM or any virtual or physical connections over which bridged Ethernet traffic is carried.

Figure 4-13 *VPLS Reference Model*

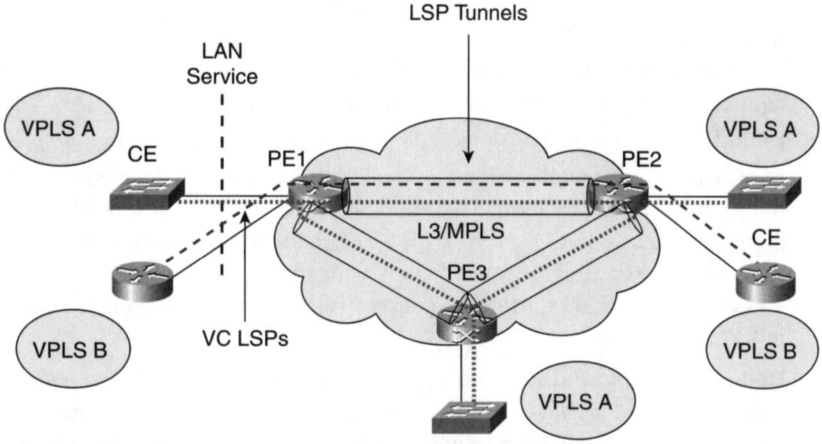

The following sections discuss the different aspects of a VPLS model:

- VPLS requirements
- Signaling the VPLS service
- VPLS encapsulation
- Creating a loop-free topology
- MAC address learning
- MAC address withdrawal
- Unqualified versus qualified learning
- Scaling the VPLS service via hierarchical VPLS (HVPLS)
- Autodiscovery
- Signaling using BGP versus LDP
- Comparing the Frame Relay and MPLS/BGP approaches
- L2VPN BGP model
- Frame Relay access with MPLS edge/core
- Decoupled Transparent LAN Service (DTLS)

VPLS Requirements

Following are the basic requirements of a VPLS service:

- **Separation between VPLS domains**—A VPLS system must distinguish different customer domains. Each customer domain emulates its own LAN. VPLS PEs must maintain a separate virtual switching instance per VPN.

- **MAC learning**—A VPLS should be capable of learning and forwarding based on MAC addresses. The VPLS looks exactly like a LAN switch to the CEs.
- **Switching**—A VPLS switch should be able to switch packets between different tunnels based on MAC addresses. The VPLS switch should also be able to work on 802.1p/q tagged and untagged Ethernet packets and should support per-VLAN functionality.
- **Flooding**—A VPLS should be able to support the flooding of packets with unknown MAC addresses as well as broadcast and multicast packets. Remember that with Ethernet, if a switch does not recognize a destination MAC address, it should flood the traffic to all ports within a certain VLAN. With the VPLS model shown in Figure 4-13, if a VPLS-capable device receives a packet from VPLS A with an unknown MAC destination address, the VPLS device should replicate the packet to all other VPLS-capable devices that participate in VPLS A.
- **Redundancy and failure recovery**—The VPLS should be able to recover from network failure to ensure high availability. The service should be restored around an alternative path, and the restoration time should be less than the time the CEs or customer L2 control protocols need to detect the failure of the VPLS. The failure recovery and redundancy of MPLS depends on how fast MPLS paths can be restored in case of a failure and how fast the network can stabilize. Chapter 6 discusses MPLS fast restoration.
- **Provider edge signaling**—In addition to manual configuration methods, VPLS should provide a way to signal between PEs to auto-configure and to inform the PEs of membership, tunneling, and other relevant parameters. Many vendors have adopted LDP as a signaling mechanism; however, there are some who prefer BGP as used in RFC 2547, *BGP/MPLS VPNs*.
- **VPLS membership discovery**—The VPLS control plane and management plane should provide methods to discover the PEs that connect CEs forming a VPLS. Different mechanisms can be used to achieve discovery. One method is via the use of BGP, as adopted in the L3VPN model. However, there is some disagreement in the industry on whether BGP implementations are appropriate, due to the complexity of BGP and the fact that it cannot signal a different label to each VPLS peer, as required by MAC learning. A proposal for using BGP promotes the use of block label distribution, as explained in the "DTLS—Decoupling L2PE and PE Functionality" section later in this chapter.
- **Interprovider connectivity**—The VPLS domain should be able to cross multiple providers, and the VPLS identification should be globally unique.
- **VPLS management and operations**—VPLS configuration, management, and monitoring are very important to the success of the VPLS service. Customer SLAs should be able to be monitored for availability, bandwidth usage, packet counts, restoration times, and so on. The metrics that have been defined by the MEF regarding performance and bandwidth parameters should apply to the VPLS service.

Signaling the VPLS Service

Signaling with VPLS is the same as described in the section "Ethernet over MPLS—Draft-Martini," with LDP using a forwarding equivalency class element. The main difference is that in the P2P martini tunnel, the VC ID is a service identifier representing a particular service on the Ethernet port, such as a different P2P VLAN. With VPLS, the VC ID represents an emulated LAN segment, and its meaning needs to be global within the same provider and across multiple providers.

VPLS Encapsulation

VPLS encapsulation is derived from the martini encapsulation used for a P2P EoMPLS service. The packet is always stripped from any service-related delimiter that is imposed by the local PE. This ensures that the Ethernet packet that traverses a VPLS is always a customer Ethernet packet. Any service delimiters, such as VLAN or MPLS labels, can be assigned locally at the ingress PE and stripped or modified in the egress PE.

Creating a Loop-Free Topology

The problem with having a VPLS domain emulate a LAN is that it can create the same circumstances that create a loop in a LAN. With L2 Ethernet networks, Spanning Tree Protocol is used to prevent loops caused by the L2 flooding mechanism. In the case of VPLS, the same scenario could happen as illustrated in Figure 4-14.

Figure 4-14 *L2 Loops*

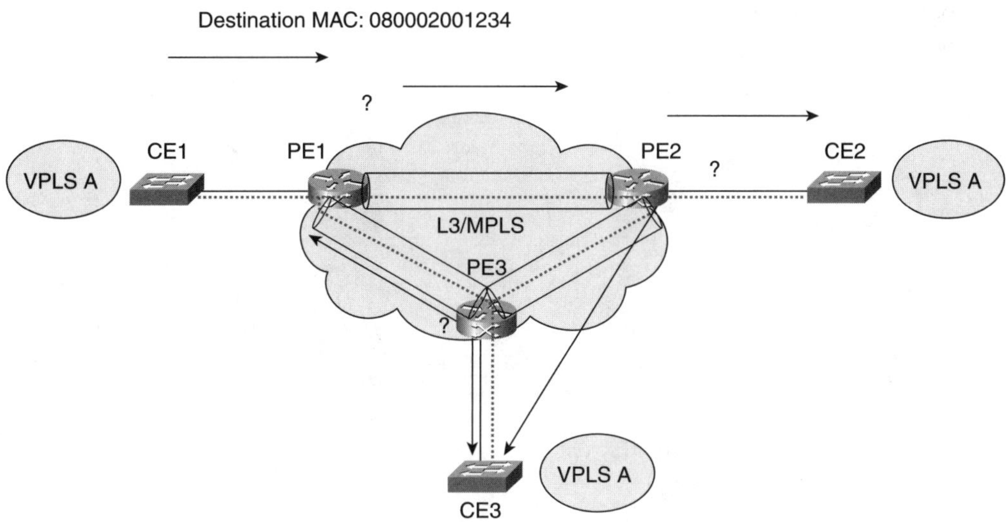

In Figure 4-14, three LSP tunnels connect PE1, PE2, and PE3. VPLS A is emulating a LAN that is carried over these LSP tunnels. If CE1 sends a packet with a destination MAC address, say 080002001234, that is unknown by PE1, PE1 has to flood, or replicate, that packet over the two tunnels connecting it to PE2 and PE3 that participate in the same VPLS. If PE2 does not know of the destination as well, it sends the packet to PE3 and CE2. In the same manner, if PE3 does not know of the destination, it sends the packet to PE1 and CE3, and the loop continues. To break the loop, Spanning Tree Protocol (STP) has to run on the PEs, and in the same way that Ethernet frames are tunneled over the MPLS LSPs, STP BPDUs also have to be tunneled.

To avoid the deployment of spanning trees, a full mesh of LSPs needs to be installed between the PEs, and each PE must support a split-horizon scheme wherein the PE must not forward traffic from one PW to another in the same VPN. This works because each PE has direct connectivity to all other PEs in the same VPN.

In Figure 4-15, a full mesh of tunnel LSPs and VC-LSPs (which are used to demultiplex the service over the tunnel LSPs) has been configured between all PEs. A PE receiving a packet over a VC-LSP cannot forward that packet to other VC-LSPs. PE1 receives an Ethernet packet with an unknown destination and replicates that packet over the three VC-LSPs that connect it to PE2, PE3, and PE4. Because there is a full mesh, PE2, PE3, and PE4 assume that the same packet they received has already been sent by the other PEs and thus do not replicate it. This prevents loops from taking place. Requiring a full mesh of LSPs becomes an issue if the PE functionality is moved closer into the access cloud, such as in the basement of multitenant unit (MTU) buildings. This would create an explosion of an LSP mesh that does not scale. The section "Scaling the VPLS Service Via Hierarchical VPLS" later in this chapter explains how such a scenario is solved.

Figure 4-15 *Avoiding Loops Via Full Mesh*

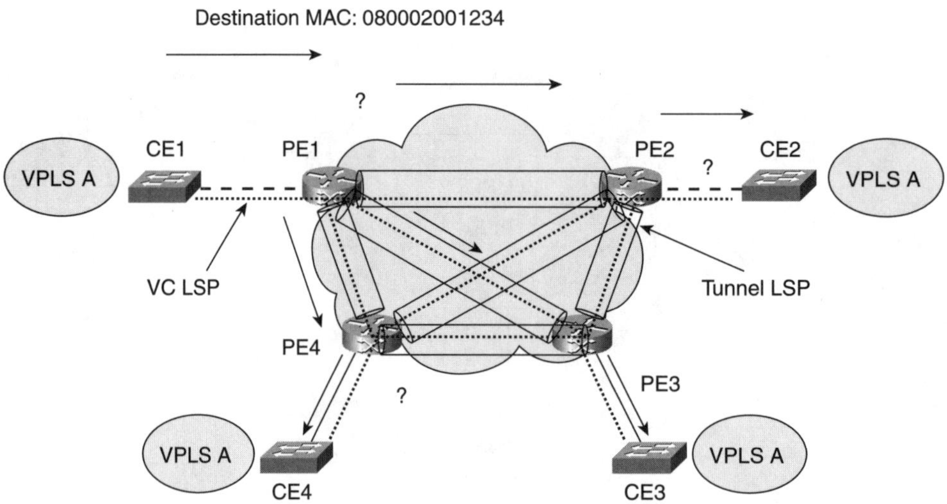

In some cases, an enterprise customer can create a backdoor loop by connecting multiple sites directly via an L2 connection. To avoid loops, STP can be run on the CEs, and the STP BPDUs are tunneled over the MPLS cloud like any other data packets. This is shown in Figure 4-16.

Figure 4-16 *Backdoor Loops*

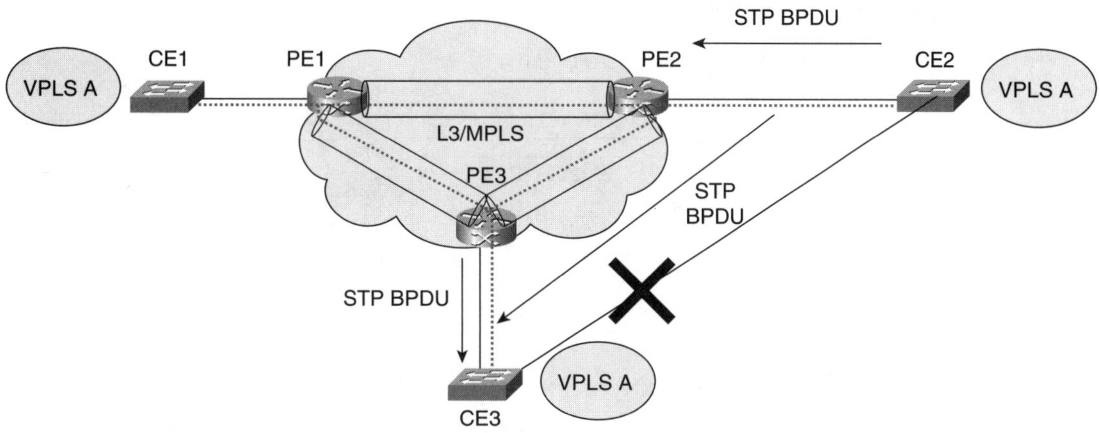

In Figure 4-16, CE2 and CE3 have a direct L2 connection, creating a loop, because traffic between CE2 and CE3 can traverse two different paths: the direct connection and the MPLS cloud. If the customer runs STP between the two CEs, STP BPDUs can be tunneled over the MPLS cloud, causing the loop to break. In case the MPLS connection or the direct connection fails, traffic is switched over the remaining connection.

So, there are many ways to create loops in an Ethernet L2 switched architecture. Some of these loops can be caused by the service provider equipment and some by the enterprise equipment itself. This creates more strain on the operation and management of these systems, when problems occur. Well-defined rules need to be set to indicate which L2 control PDUs are to be carried over the provider network. This prevents finger-pointing between the customer and provider in case problems such as broadcast storms occur.

MAC Address Learning

In the P2P PW scenarios in which Ethernet packets come in on one side of the PW and come out on the other side, MAC learning is not necessary. What goes in the tunnel comes out on the other end. VPLS operates in any-to-any MP2MP mode. This means that a PE is connected with multiple VC-LSPs to different PEs that participate in the multiple VPLS domains, and the PE needs to decide which LSPs to put the traffic on. This decision is based on destination MAC addresses that belong to a certain VPLS. It is unreasonable to assume that this function can be statically configured (although it could), because many MAC addresses would need to be mapped to many LSPs. MAC learning allows the PE to determine from which physical port or LSP a particular MAC address came.

Figure 4-17 shows an example of how MAC learning is achieved.

Figure 4-17 *MAC Learning*

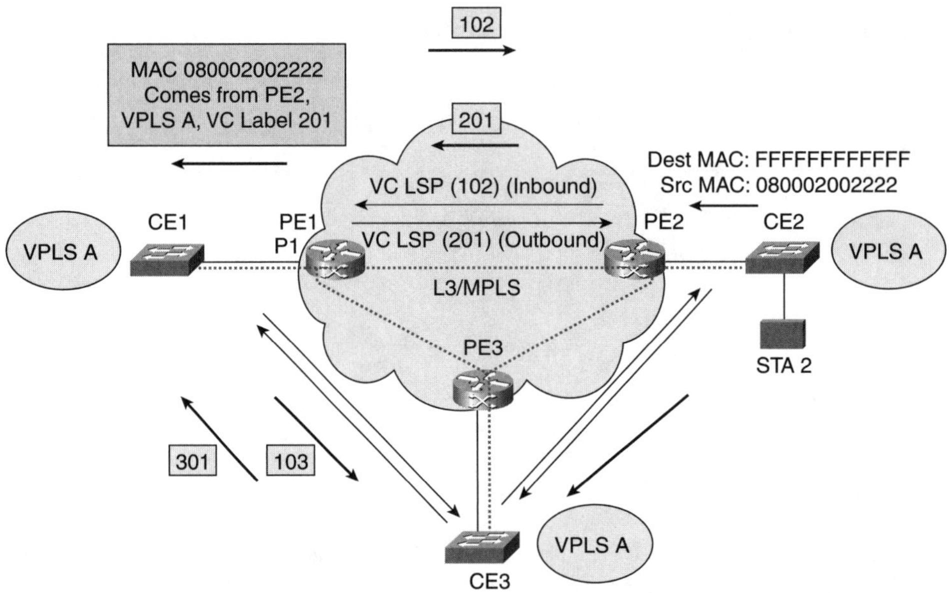

In Figure 4-17, PE1, PE2, and PE3 establish pairs of VC-LSPs between each other as follows:

1 Using LDP, PE2 signals VC label 201 to PE1, and PE1 signals VC label 102 to PE2.

2 A station behind CE2, STA 2, with a MAC address of 080002002222, sends a broadcast packet to destination MAC FFFFFFFFFFFF. PE2 recognizes that STA 2 belongs to VPLS A (via configuration or other mechanism) and replicates the packet to the two VC-LSPs connected to PE1 and PE3 that also participate in VPLS A.

3 When the packet comes to PE1 on PE1's inbound VC-LSP, it associates the MAC address of STA 2 with the "outbound" VC-LSP in the same VC-LSP pair that constitutes the PW between PE2 and PE1.

4 From then on, if PE1 receives a packet destined for MAC 080002002222, it automatically sends it on its outbound VC-LSP using label 201 (which was signaled by PE2).

This process constitutes MAC learning on the VC-label side. Standard Ethernet MAC learning occurs on the Ethernet port side, where PE2, for example, associates MAC 080002002222 with its local Ethernet port or the VLAN it came on. This process continues until all PEs have learned all MAC addresses on their local ports/VLANs and across the MPLS cloud. Notice that PE1 signals two different labels to PE2 and PE3. In this example, PE1 signals label 102 to PE2 and 103 to PE3. This way, PE1 can distinguish inbound packets from PE2 and PE3.

MAC Address Withdrawal

L2 Ethernet switching includes a mechanism called MAC aging that lets MAC addresses be aged out of an Ethernet switch MAC table after a certain period of inactivity. In some cases, such as an MTU building that is dual-homed to two different Ethernet switches in the central office (CO), faster convergence can occur if a mechanism exists to age out (withdraw) or relearn MAC addresses in a way that is faster than the traditional L2 MAC aging. The IETF has defined a MAC type length value (TLV) field that can be used to expedite learning of MAC addresses as a result of topology change.

Unqualified Versus Qualified Learning

When a PE learns MAC addresses from the attached customers, these MAC addresses are kept in a Forwarding Information Base (FIB). The FIB should keep track of the MAC addresses and on which PWs they were learned. This allows MAC addresses to be tracked by VPLS. This is different from the traditional MAC learning of Ethernet switches, where all MAC addresses are shared by a single customer. VPLS can operate in two learning modes, unqualified and qualified.

In unqualified learning, a customer VPLS is a port-based service where the VPLS is considered a single broadcast domain that contains all the VLANs that belong to the same customer. In this case, a single customer is handled with a single VPLS. On the other hand, qualified learning assumes a VLAN-based VPLS where each customer VLAN can be treated as a separate VPLS and as a separate broadcast domain. The advantage of qualified learning is that customer broadcast is confined to a particular VLAN.

Scaling the VPLS Service Via Hierarchical VPLS

The VPLS service requires a full mesh of VC-LSPs between the PE routers. This works adequately if the PE routers are contained in COs and the different customers are aggregated in these COs. In the case of MTU deployments, the PEs are deployed in the building basements where multiple customers are aggregated. In this case, starting the VPLS service in the PE might cause scalability problems because there are many more buildings than COs. A full mesh of LSPs between all the buildings that participate in the VPLS service would cause an unmanageable LSP deployment. For x PEs that are deployed, $x * (x - 1) / 2$ bidirectional LSPs need to be deployed. Remember also that it takes two LSPs—one inbound and one outbound—to construct a bidirectional PW, which means that $x * (x - 1)$ unidirectional VC-LSPs need to be signaled.

Figure 4-18 shows a deployment in which the VPLS starts in the basement of MTU buildings and a full mesh of LSPs is required between PEs. This LSP explosion will cause an operational nightmare.

For any "new" MTU building that is added to the network, the new MTU must be meshed to every PE in the existing MTUs, which doesn't scale. Packets get flooded over all LSPs participating in a VPLS; if the MAC destination is unknown, this puts a big load on the MTU PE.

Figure 4-18 *Full Mesh*

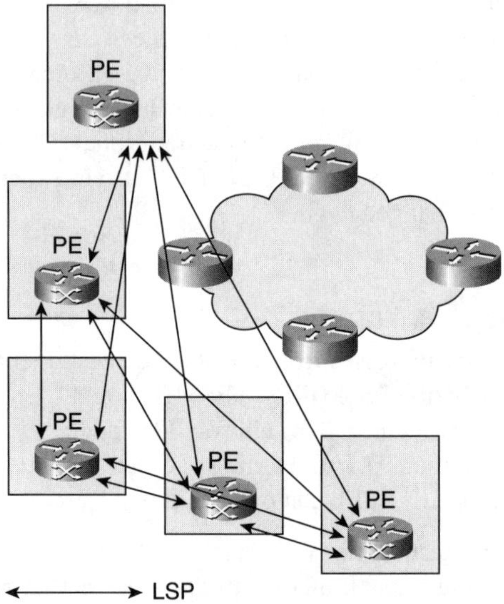

A better approach for MTUs is to create a hierarchical VPLS (HVPLS) model in which the MTU PEs establish access tunnels (spokes) to the CO PEs, and the CO PEs (hubs) establish a full mesh. This is shown in Figure 4-19.

Figure 4-19 *HVPLS*

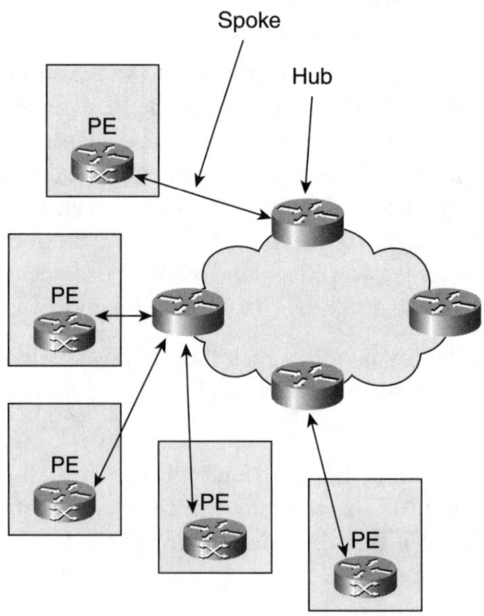

The hierarchical model scales better, because a new MTU that is added to the network has to establish an LSP only with the local PE and does not need to establish LSPs with every other PE. This is a major operational cost saving.

There are many flavors for the MTU and the CO PEs. The IETF has adopted the following terminology that is used in the rest of the chapter:

- **MTU-s**—This is a PE that is placed in the MTU and is capable of doing MAC learning and L2 switching/bridging. This could be a pure L2 Ethernet switch, or an L2 Ethernet switch that is capable of MPLS tagging and forwarding but does not have to do any IP routing.
- **PE-r**—This is a PE that is capable of IP routing/MPLS but is not capable of MAC learning. This device can be placed in the MTU or the CO. This is basically an IP/MPLS router.
- **PE-rs**—This a PE that is capable of both L2 switching and IP routing.

The following sections explain two different scenarios used in service provider deployments: one for MTU-s deployments and the other for PE-rs.

MTU Device Supports MAC Learning and L2 Switching (MTU-s)

In this scenario, the MTU-s is an L2 Ethernet switch that is capable of MAC learning and can do switching based on MAC addresses. The MTU-s does all the normal bridging functions of learning and replications on all its ports, including the virtual spoke ports, which are the PWs that connect the MTU-s to the PE-rs. The ability of the MTU-s to do MAC learning and bridging simplifies the signaling between the MTU-s and the PE-rs at the CO, because the MTU-s can associate all the access ports belonging to the same VPLS with a single PW between the MTU-s and the PE-rs. This is better illustrated in Figure 4-20.

Figure 4-20 *Sample MTU-s*

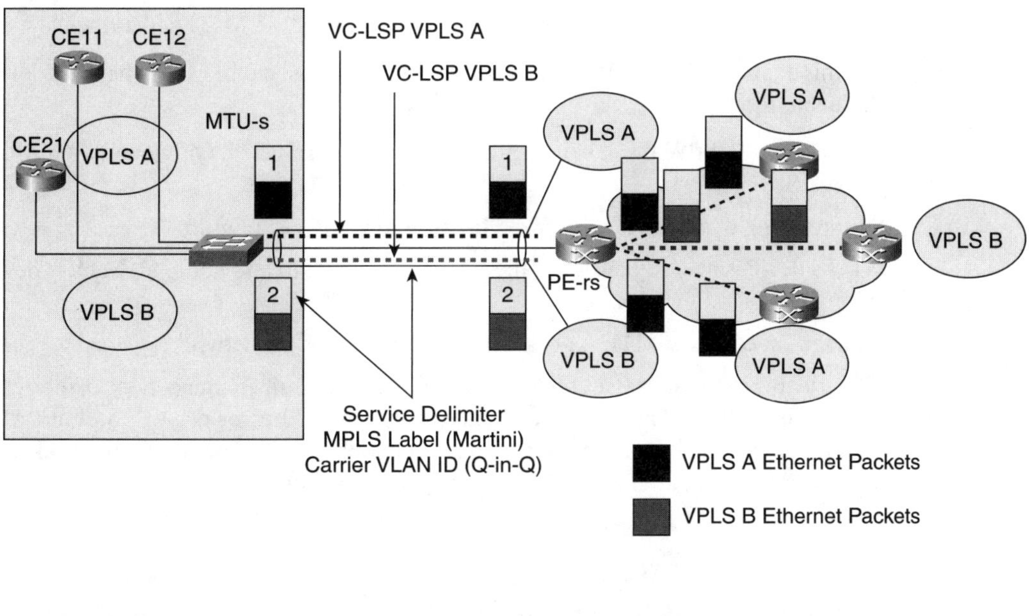

In Figure 4-20, a service provider is offering a VPLS service to two customers via an MTU-s in the basement of the building. The MTU-s connects to a PE-rs in the CO. Note that on the MTU-s, two access Ethernet ports are assigned to VPLS A. These access ports are connected to CE11 and CE12, which both connect to the same customer. This scenario could occur if the same customer has two different locations that are in the same MTU or a nearby building and all connections are serviced by the same MTU-s toward the CO. In this case, the MTU-s switches all traffic between CE11 and CE12 locally via regular L2 switching and switches traffic between CE11 and CE12 and the remote sites in VPLS A using a single PW between the MTU-s and the PE-rs in the CO.

Because the MTU-s also services VPLS B, the service provider has to assign a service delimiter to traffic coming from VPLS A and VPLS B to differentiate between the two customers. Remember that this was discussed in Chapter 3, "Metro Ethernet Services," which introduced the concept of using carrier VLAN IDs to differentiate between customer traffic. In this example, you could set up the spoke in two ways:

- The service provider is using Q-in-Q to separate the customer traffic on the MTU-s and to indicate to the PE-rs which traffic belongs to which VPLS. In this case, the service delimiter is a carrier VLAN ID carried on top of the customer's Ethernet packet. The customer traffic itself also could carry customer-specific VLAN tags; however, those tags are not seen by the service provider.
- The service provider is using two martini EoMPLS PWs to carry traffic from the different customers. In this case, the MPLS tag on top of the customer's Ethernet traffic is the service delimiter recognized by the PE-rs.

The decision of whether to use Q-in-Q or martini tunnels depends on the equipment the vendor uses in the MTU and the CO. In some cases, the MTU equipment doesn't support MPLS. In other cases, the MTU and CO equipment does not interoperate when using Q-in-Q. You should also remember that some Ethernet switch vendors support neither VLAN stacking on a per-customer basis nor MPLS. You should not use such equipment in MTU deployments.

Notice in this example that the PWs used between the MTU-s and the PE-rs have achieved multiple functions:

- The need for full PW mesh between the MTU-rs is eliminated. Only one PW is used per VPLS.
- The signaling overhead is minimized because fewer PWs are used.
- MTU-s devices are only aware of the PE-rs they attach to and not to all MTU-s devices that participate in the VPLS.
- An addition of a new MTU-s does not affect the rest of the network.

The MTU-s learns MAC addresses both from the Ethernet customer connections in the building and from the spoke PWs. The MTU-s associates the MAC addresses per VPLS. If an MTU-s receives a broadcast packet or a packet with an unknown destination MAC, the packet is flooded (replicated) over all the MTU-s physical or logical connections that participate within the VPLS. Note that there is one PW per VPLS on the spoke connection, so the packet is replicated only once per VPLS.

The MTU-s device and the PE-rs device treat each spoke connection like a physical port on the VPLS service. On the physical ports, the combination of the physical port and VLAN tag is used to associate the traffic with a VPLS instance. On the spoke port, the VC label or carrier VLAN ID (for Q-in-Q) is used to associate the traffic with a particular VPLS. L2 MAC address lookup is then used to find out which physical port the traffic needs to be sent on.

The PE-rs forms a full mesh of tunnels and PWs with all other PE-rs devices that are participating in the VPLS. A broadcast/multicast or a packet with an unknown MAC destination is replicated on all PWs connected to the PE-rs for a certain VPLS. Note that the PE-rs can contain more VPLS instances than the MTU-s, because the PE-rs participates in all the VPLSs of the MTU buildings that are attached to it, while the MTU-s only participates in the VPLS of the customers in a particular building. Also, the MAC learning function is done twice: once at the MTU-s and another time at the PE-rs.

PE-rs Issues with MAC Learning

The fact that the PE-rs is doing MAC learning raises concerns with service providers. The PE-rs has to learn all the MAC addresses that exist in all VPLS instances it participates in. This could be in the hundreds of thousands of MAC addresses that need to be learned if the VPLS service is delivering LAN connectivity between CEs that are L2 switches. Remember that a VPLS emulates a LAN service and learns all MAC addresses it hears from all stations connected to the LAN. If the CEs are L2 Ethernet switches, the VPLS will learn all MAC addresses behind the Ethernet switch. Some of these concerns can be alleviated through different approaches:

- If the CE equipment is an IP router, the VPLS learns only the MAC addresses of the IP router interfaces that are connected to the VPLS. MAC stations behind IP routers are hidden, because IP routers route based on IP addresses and not MAC addresses. In this model, the MAC address space is very manageable.
- If the CEs are L2 switches, it is possible to use filtering mechanisms on the MTU-s to allow service for only a block of the customer's MAC addresses and not all of them. Filtering helps reduce the explosion of MAC addresses on the PE-rs; however, it adds more management overhead for both the customer and the service provider.

A different model can be used to allow the MTU-s to do MAC learning at the building and not to do MAC learning at the PE-rs. This model is called the *Decoupled Transparent LAN Service (DTLS),* which is explained later in this chapter in the section "DTLS—Decoupling L2PE and PE Functionality."

Non-Bridging Devices as Spokes

In some cases, existing IP routers are deployed as spokes. As previously described, the IETF calls such a device a PE-r, to indicate routing functionality only. These routers are not capable of bridging and cannot switch packets based on MAC addresses. To offer an L2 service using the PE-r, it is possible to create PWs between the PE-r and the CO PE-rs, where all the L2 switching functions are done at the CO. This model creates more overhead, because unlike the MTU-s, where all access ports belonging to the same VPLS are mapped to a single PW, the PE-r requires that each access port is mapped to its own PW. This is illustrated in Figure 4-21.

Figure 4-21 *Spoke Device Is a Router*

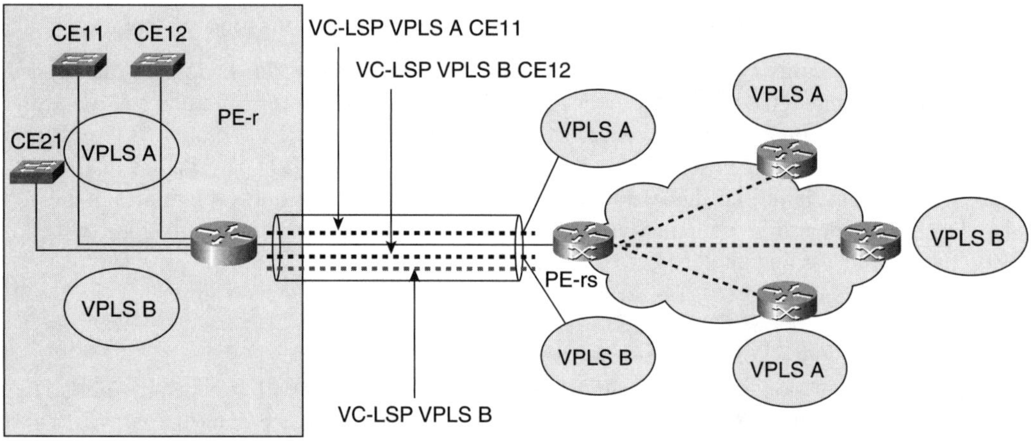

Figure 4-21 uses a PE-r as a spoke. Note that VPLS A now requires two PWs—one for CE11 and one for CE12—that belong to the same customer. For any traffic that needs to be switched between the two access ports of the same customer that are connected to CE11 and CE12, that traffic needs to be transported to the CO and switched at the PE-rs.

Dual-Homed MTU Devices

It is possible to dual-home an MTU device to protect against the failure of a spoke or the failure of a PE-rs at the CO. *Dual-home* refers to connecting the MTU device via two separate spokes.

Figure 4-22 shows an MTU-s device that is dual-homed to the PR-rs at the CO via two PWs, one primary and one backup. To prevent an L2 loop in the network, the primary PW is active and passing traffic while the secondary PW is inactive. In this scenario, spanning tree is not needed, because only a single PW is active at the same time. In normal operation, all PE-rs devices participating in VPLS A learn the MAC addresses behind MTU-s via the primary PW connected to PE1-rs. The following two scenarios might take place:

- **Failure of the primary PW**—In this case the MTU-s immediately switches to the secondary PW. At this point the PE2-rs that is terminating the secondary PW starts learning MAC addresses on the spoke PW. The speed of convergence in the network depends on whether MAC TLVs are used, as described in the "MAC Address Withdrawal" section earlier in this chapter. If the MAC address TLVs are used, PE2-rs sends a flush message to all other PE-rs devices participating in the VPLS service. As such, all PE-rs devices converge on PE2-rs to learn the MAC addresses. If the MAC TLV is not used, the network is still operational and converges using the traditional L2 MAC learning and aging. During this slow convergence, the PE-rs devices slowly learn the MAC addresses in the network.

- **Failure of the PE1-rs**—In this case, all PWs that are terminated at PE1-rs fail, and the network converges toward PE2-rs.

Figure 4-22 *Dual-Homed MTU Device*

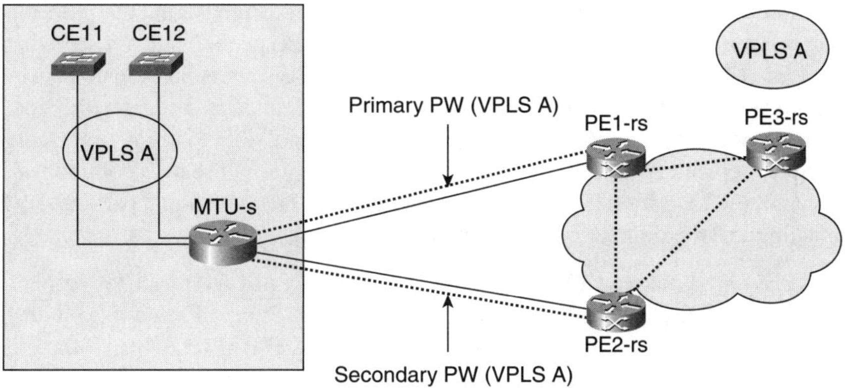

Autodiscovery

Autodiscovery refers to the process of finding all the PEs that participate in a given VPLS. So far, this chapter has assumed that this function is manual, meaning that the network operator dedicates certain PEs to belong to a certain VPLS and configures that information on each PE belonging to the VPLS. This process can be configuration-intensive, especially with a large number of PEs, because manual configuration and deletion are needed every time a PE is added to or removed from the network. With autodiscovery, each PE discovers which other PEs are part of the same VPLS and discovers PEs when they are added to or removed from the network. Different mechanisms have been proposed by different vendors, such as the use of BGP, LDP, or DNS to achieve autodiscovery. This section elaborates on BGP and how it compares with LDP.

BGP uses the concept of extended communities to identify a VPLS. PEs exchange information via direct Internal BGP (IBGP) or External BGP (EBGP) peering or route reflectors.

You saw at the beginning of this chapter that BGP is used with MPLS L3VPNs to achieve discovery of VPN information. A similar approach is used to achieve VPLS autodiscovery by having the routes exchanged in BGP carry a VPN-L2 address. A VPN-L2 address contains a route distinguisher (RD) field that distinguishes between different VPN-L2 addresses. Also, a BGP route target (RT) extended community is used to constrain route distribution between PEs. The RT is indicative of a particular VPLS. Because a PE is fully meshed with all other PEs, it receives BGP information from all PEs. The PE filters out the information based on the route target and learns only information pertinent to the route targets (VPLSs) it belongs to.

Signaling Using BGP Versus LDP

In this chapter, you have learned about the use of LDP as a signaling mechanism to establish and tear down PWs between PEs. Some vendors have adopted BGP as a signaling mechanism because of its scalability and its ability to support VPLS deployment across multiple providers. This section presents a more detailed comparison of the use of LDP and BGP as a signaling mechanism and BGP.

With LDP used as the signaling protocol, targeted LDP sessions are established between PE peers. An LDP session is called "targeted" because it is set directly between two PEs that do not have to be adjacent. These PEs exchange MPLS labels directly, and that information is hidden from the routers that exist on the path between these PE peers. You have seen that a full mesh of these peers between PEs is needed per VPLS. If all PE routers participate in every VPLS, a full mesh is needed between all PEs. Also, each PE needs to carry a separate FIB per VPLS, which increases the number of FIBs per PE. However, it is possible to segment the network into PEs that have separate VPLS coverage, meaning that they do not serve a common set of VPLSs. In this case, the LDP mesh is needed only between the PEs covering a particular VPLS, and the signaling and the number of FIBs per PE are reduced.

If all PEs participate in all VPLS instances, there is a full LDP mesh between all PEs, and each PE carries a FIB per VPLS, as shown in Figure 4-23, Part A. Figure 4-23, Part B, shows that three PEs participate in VPLS A and carry a VPLS A FIB (FIB A) while the other PEs carry a VPLS B FIB. Note that a full mesh between all PEs is not required.

Figure 4-23 *LDP Signaling Options*

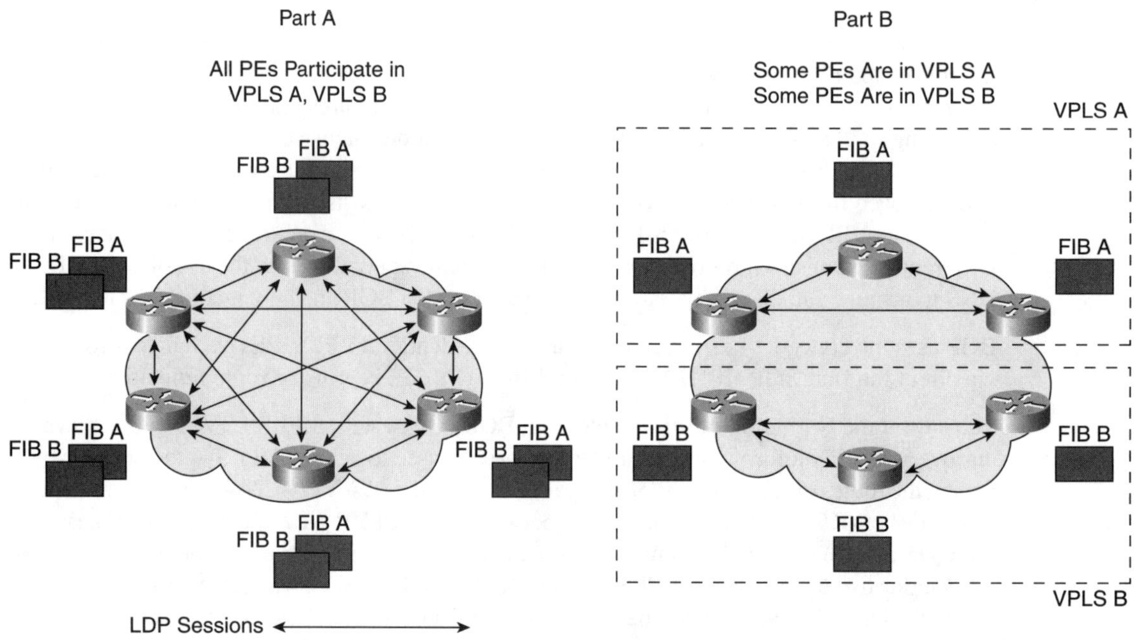

Vendors proposing BGP as a signaling mechanism between PEs argue that BGP offers more scalability and is already proven to work for L3VPNs as defined in RFC 2547. Also, BGP can be used for both signaling and PE discovery, whereas LDP is used only for signaling. BGP uses what is called a *route reflector* to solve the full-mesh PE-to-PE session issue and the fact that, with LDP, every time a new PE is added to the network, a full mesh needs to be established with all PEs (in the same VPLS). The route reflector concept allows PEs to operate in a client/server model, where the PEs peer with a single or multiple route reflectors (for redundancy), and the route reflector relays information between the different PEs. In this case, if a new PE is added to the network, that PE needs to establish only a single peering session with the route reflector.

Figure 4-24 shows all PEs peering with a route reflector. A new PE added to the network has to peer only with the route reflector.

Figure 4-24 *Signaling Via BGP with Route Reflectors*

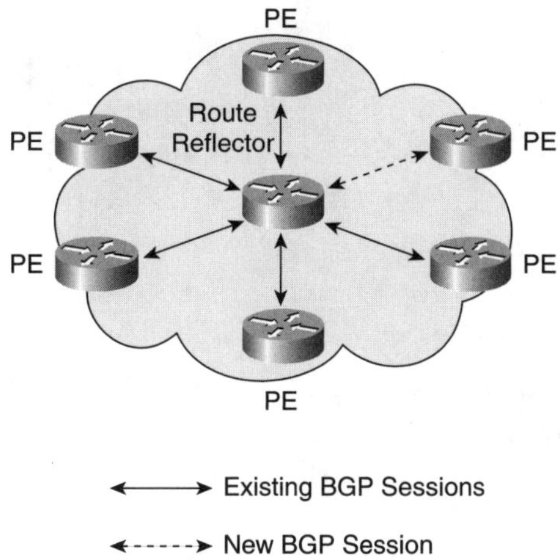

On the other hand, using BGP does create the issue of requiring label ranges, because BGP cannot direct label mappings to a specific peer. The use of label ranges is covered in the upcoming section "L2VPN BGP Model."

Comparison Between the Frame Relay and MPLS/BGP Approaches

This section first briefly compares Frame Relay VPNs and MPLS L2VPNs and then delves into a discussion about how some IETF drafts proposing BGP are influenced by the Frame Relay VPN model.

As Figure 4-25 shows, a Frame Relay VPN with any-to-any connectivity between the different sites requires a full mesh of PVCs between the different CEs. The network uses Frame Relay at the access and Frame Relay/ATM at the edge and core. The physical connection between the CE and the Frame Relay network is assigned multiple DLCIs, and each DLCI is used to switch the traffic from one CE all the way to another CE.

Figure 4-25 *Frame Relay Access, Edge, and Core*

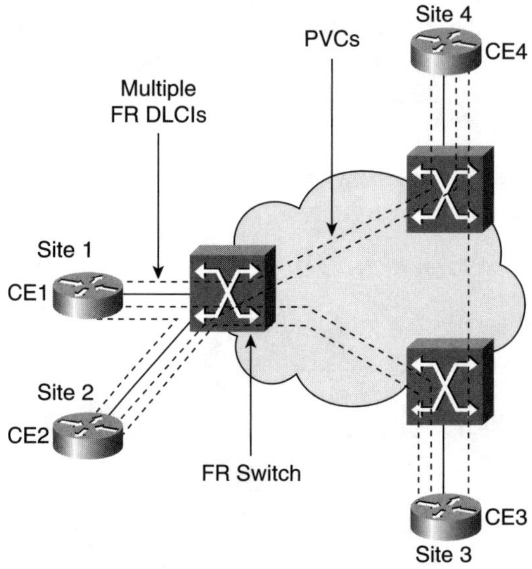

Such networks have two drawbacks. First, the whole network is locked into a single technology, such as Frame Relay or ATM. Second, adding a new site into the VPN and connecting that site to the rest of the VPN causes an operational headache because many PVCs need to be configured site to site.

Figure 4-26 shows the same VPN but with an MPLS deployment at the core and with the possibility of using Frame Relay, Ethernet, or MPLS at the access.

On the physical connection between the CE and PE, Frame Relay DLCI can still be used to indicate the particular service. However, these services are now carried through pre-established packet tunnels in the network. The provisioning is now simplified, because rather than configuring end-to-end PVCs in the network to establish connectivity between the different sites, the same can be accomplished by assigning the right DLCIs at the CE-to-PE connection; however, the services to different CEs are carried over pre-established tunnels in the network.

L2VPN BGP Model

The L2VPN BGP model introduces some new terminology for referencing customer sites, customer equipment, and the way blocks of MPLS labels are allocated in the network.

Figure 4-26 *Frame Relay Access and MPLS Edge/Core*

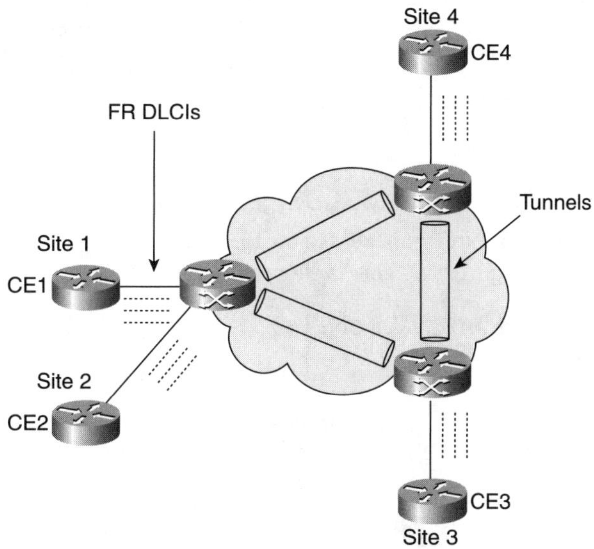

The L2VPN BGP model divides the network into two levels:

- **The provider backbone**—Contains all the PEs.
- **The sites**—These are the different locations where the customer equipment (CE) resides. A site can belong to a single customer and can have one or more CEs. Each CE is referenced using its own CE ID that is unique within the VPLS.

  In other scenarios, a site can belong to multiple customers, in the case of an MTU, and each customer can have one or more CEs in that site. In this case, a customer connection is represented via a combination site ID and VPLS ID and a physical port on the MTU device.

In the BGP L2VPN scheme, each PE transmits pieces of information such as label blocks and information about the CEs to which it connects to all other PEs. To reach a destination, the PE need only install a route to the site where the destination exits. This allows the service to scale well, because this model tracks the number of VPN sites rather than individual customers.

The L2VPN BGP model is generalized to cover the following:

- **Connectivity of a CE to a PE**—In this model, the CE is a Frame Relay–capable or MPLS-capable device and can allocate a Frame Relay DLCI or an MPLS label after negotiation with the PE.
- **Connectivity of a CE to an L2PE and then to a PE**—This model reflects an MTU installation where an L2PE is used to connect the multiple customers within a site to a common piece of equipment in the basement. The L2PE is similar to the MTU-s that was already discussed. The L2PE is a switch that does MAC learning and bridging/switching,

and it encapsulates the Ethernet customer traffic inside an MPLS packet with labels that are allocated by the PE. In the MTU case, the CE need only have an Ethernet connection to the L2PE and does not need to have any MPLS functionality.

The following section describes an example of an L2VPN with Frame Relay connectivity used on the access and MPLS used on the edge/core.

Example of Frame Relay Access with MPLS Edge/Core

In this scenario, the CEs are connected to the PEs via Frame Relay, and the PEs carry the service over the network using MPLS. This is shown in Figure 4-27.

Figure 4-27 *Frame Relay Access with MPLS Edge/Core*

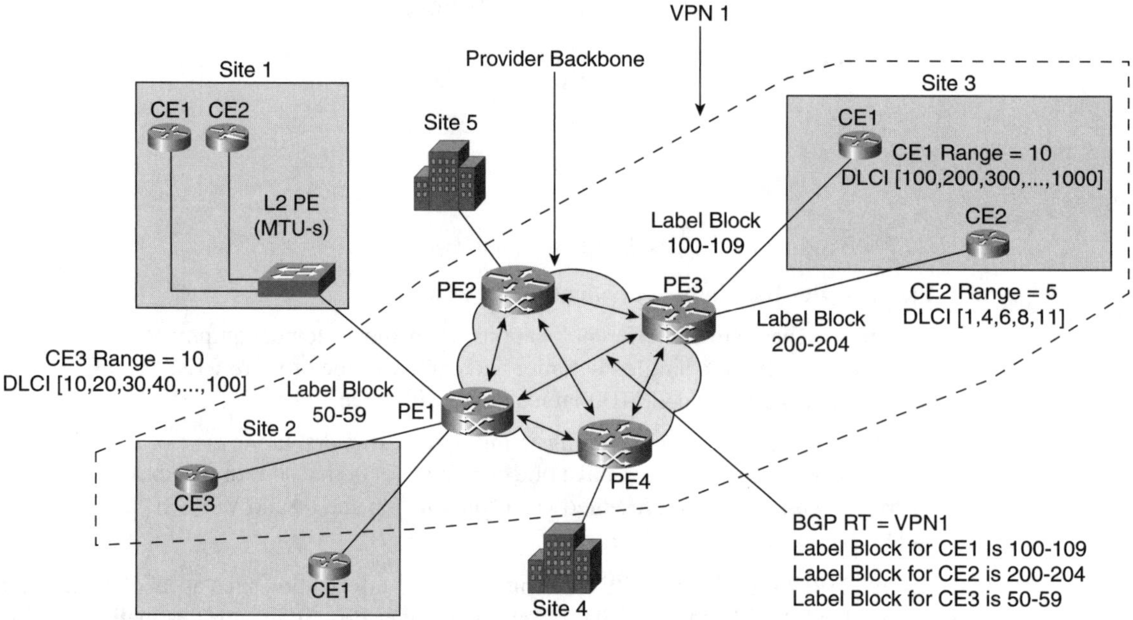

Both CE and PE must agree on the FR DLCI that will be used on the interface connecting them. Each CE that belongs to a VPN is given a CE ID, which is unique in the context of the VPN. In Figure 4-27, a VPN consists of the three CEs: CE1, CE2, and CE3, where CE1 and CE2 are located in site 3 and CE3 is located in site 2. The CE IDs 1, 2, and 3 are supposed to be unique within the same VPN.

Each CE is configured with a list of Frame Relay DLCIs that allows it to connect to the other CEs in the VPN. The size of this list for a particular CE is called the CE's range. In Figure 4-27, for CE3 in site 2 to connect to both CE1 and CE2 in site 3, it would need two DLCIs, one for each remote CE. As such, the CE range determines the number of remote sites a CE can connect to. The larger the range, the more remote CEs a CE can connect to. Each CE also knows which DLCI connects it to every other CE. When a packet comes to a CE from inside the customer network, the CE can use the correct DLCI based on where that packet is going. From then on, the packet is "switched" from one end to the other. The network behaves as a Frame Relay switch with respect to the CEs.

Each PE is configured with the VPNs in which it participates. For each VPN, the PE has a list of CEs that are members of that VPN. For each CE, the PE knows the CE ID, the CE range, and which DLCIs to expect from the CE. When a PE is configured with all the needed information for a CE, it chooses an MPLS label block, which is a contiguous set of labels. The number of these labels is the initial CE range, meaning if the CE has a range of ten DLCIs, the PE chooses ten MPLS labels. The smallest label in this label block is called the *label base,* and the number of labels in the label block is called a *label range.* The PE then uses BGP Network Layer Reachability Information (NLRI) to advertise the label blocks and the CE ID to which it connects, to all other PEs. Only the PEs that are part of the VPN (through the use of the BGP route target) accept the information. Other PEs discard the information or keep it for future use if they become part of that VPN.

A CE can have one or more label blocks, because when the VPN grows, more CEs participate in the VPN, and the CE label ranges might need to be expanded. If a CE has more than one label block, the notion of block offset is used. The *block offset* identifies the position of a label block in the set of label blocks of a given CE.

In reference to Figure 4-27, PEs 1, 2, and 3 participate in VPN1. The following is an example of the information that needs to be configured on PE1 and the information that PE1 advertises and learns via BGP.

Figure 4-27 shows the following:

- CE3 in site 2 and CE1 and CE2 in site 3 all belong to VPN1, as shown by the dotted line.
- CE3 is given the following set of DLCIs: [10,20,30,40,50,60,70,80,90,100], which correspond to a CE range of 10 (10 DLCIs).
- PE1 is given an MPLS label block that contains 10 labels, from label 50 to label 59. Label range = 10, label base = 50, block offset = 0.
- CE1 is given the following set of DLCIs: [100,200,300,400,500,600,700,800,900,1000], which correspond to a CE range of 10 (10 DLCIs).
- PE3's label block for CE1 contains 10 labels, from label 100 to label 109. Label range = 10, label base = 100, block offset = 0.
- CE2 is given the following set of DLCIs: [1,4,6,8,11], which correspond to a CE range of 5 (5 DLCIs).
- PE3's label block for CE2 contains 5 labels, from label 200 to label 204. Label range = 5, label base = 1, block offset = 0.

For PE1, the following takes place:

1. PE1 is configured as part of VPN1. BGP route target = VPN1.
2. PE1 is configured to have CE3 be part of VPN1. This can be done by configuring a physical port or a combination physical port and VLAN to be part of VPN1. CE1 is then assigned to that port/VLAN.
3. PE1 learns of CE1 and CE2 and the respective label blocks' offset and label base via BGP NLRI.
4. The following label information is configured on PE1 for CE3:
   - Label block: 50–59
   - Label base = 50
   - Label range (size of the block) = 10
   - Block offset = 0 (there is only one block)

     Note that PE1's label block is the same size as CE3's range of DLCIs, which is [10, 20, 30, 40, . . . , 100].
   - PE1 advertises the ID of CE3 and the label block to all other PEs via BGP NLRI.

The choice of assigning a DLCI to a particular CE is a local matter. Some simple algorithms could be used such that the CE ID of the remote CE becomes an index into the DLCI list of the local CE (with index 0 being the first entry in the list, 1 being the second entry in the list, and so on). So, for a connection between CE3 and CE1, CE3 could be allocated the second DLCI in the list (DLCI 20) because the remote CE is CE1, and CE1 is allocated the fourth DLCI in its list (DLCI 400) because the remote CE is CE3.

The PE in turn can use a simple algorithm to identify which MPLS label is used to reach a remote CE. In our example, suppose PE1 receives a BGP NLRI from PE3, indicating that CE1 has a label block 100–109. PE1 could use the CE ID of CE3 as an index into CE1's label block. In this case, PE1 could use label 103, which is CE1's label base (100) + CE3's ID (3). PE3 then uses label 51 (CE3's label base + 1) to reach CE3. As such, a packet coming from CE3 on DLCI 20 is encapsulated with MPLS label 103, and a packet coming from CE1 on DLCI 400 is encapsulated with MPLS label 51. An additional label is used on top of the stack to indicate the PE-to-PE LSP tunnel between PE1 and PE3.

DTLS—Decoupling L2PE and PE Functionality

The concept discussed in the preceding section is extended to address the Ethernet-to-MPLS scenario—specifically, for MTU deployments. In an MTU scenario, multiple customers in the building are connected to a basement box, the MTU-s, which is referred to as "L2PE" in this section. In this case, the CEs are talking Ethernet to the L2PE, and the L2PE can talk either Ethernet or MPLS to the PE. Unlike the Frame Relay service, in which all the connections are P2P, the Ethernet service allows P2P and MP2MP VPLS service. In a multipoint service,

MAC addresses are used to distinguish how the traffic is directed over the MPLS network. MAC learning in PEs can cause scalability issues, depending on the L2 service. It is possible to decouple the functions needed to offer a VPLS service between the L2PE and the PE. These functions consist of the following:

- **MAC learning**—Learning MAC addresses from customers in the MTU and from other L2PEs across the metro
- **STP**—Building a loop-free topology on both the LAN side and the metro side
- **Discovery**—Discovering other L2PEs connected to the metro

It is possible to have the L2PE do the MAC learning and STP functions and to have the PE do the discovery function. As you know, the discovery function can be done via a protocol such as BGP to exchange information between the PEs. The benefit of this decoupling, called Decoupled Transparent LAN Service (DTLS), is to alleviate the PEs from the L2 functionality. Most PEs that have been deployed in provider networks are IP routers. These routers have been designed for IP core routing and L3 edge functionality and lack most of the functionality of L2 switches. L2 switches, on the other hand, come from an enterprise background and lack most of the scalability functions offered by L3 IP/MPLS routers.

Although new equipment is coming on the market that does both L2 switching and L3 IP routing, most of the deployed equipment are routers. DTLS allows the PE to function as IP routers and MPLS switches and puts all the L2 functionality in the L2PE. The PE does the VPN/VPLS discovery and runs BGP. The L2PE could then be a simpler L2 switch. The L2PE does not have to do any IP routing or run complex protocols such as BGP. The L2PE needs to be able to do L2 functions, such as MAC learning, STP, and bridging/switching, and be able to label the packets via either VLAN IDs or MPLS labels.

Alleviating PEs from MAC Learning in the DTLS Model

One of the main issues for L2VPN services in the metro is MAC address learning. You have seen in this book that if the service offers a LAN connection between different sites and if the CEs are L2 switches (not routers), all MAC addresses that exist in the different connected LANs become part of the VPLS. As an example, assume the following scenario in a hypothetical metro:

- The metro contains 60 PEs
- Each PE is connected to ten buildings—that is, ten L2PEs
- Each L2PE services ten customers
- Each customer has two VPLS
- Each VPLS has 100 stations

The following calculations show the difference between starting the VPLS service at the L2PE and starting the service at the PE, based on the preceding information.

If you start the VPLS service at the L2PE, each L2PE has to support the following:

- Ten customers
- 10 * 2 = 20 VPLS
- 20 * 100 = 2000 MAC addresses

Also, because there are 60 * 10 = 600 L2PEs, assuming that each L2PE talks to every other L2PE in a full mesh (BGP or LDP), the number of bidirectional sessions between the L2PEs is 600 * (600 – 1) / 2 = 179,700 sessions.

Starting the VPLS service at the PE, the PE has to support the following:

- 10 * 10 = 100 customers
- 100 * 2 = 200 VPLS
- 200 * 100 = 20,000 MAC addresses

If there is a full LDP/BGP mesh between the PEs, the number of bidirectional sessions is 60 * (60 – 1) / 2 = 1770 sessions.

From the previous calculations, the following can be easily deduced:

- Doing MAC learning at the L2PE and *not* at the PE scales much better. Otherwise, the PE has to deal with an explosion of MAC addresses.
- Doing a hierarchy in which the full mesh of BGP/LDP sessions starts at the PE prevents a session explosion.

The DTLS model keeps the MAC learning at the L2PE, assigns MPLS labels at the L2PE, and puts the VPLS discovery with BGP or other protocols at the shoulder of the PE. This way, the model can scale much better. The following needs to happen on the L2PE and on the PE:

- The L2PE:
  - Needs to behave as a bridge/switch. It should be able to learn MAC addresses from the building customers and from other L2PEs.
  - Should be able to send or receive tagged packets. The L2PE should be able to perform tag stacking and swapping and handle both VLAN and MPLS tags.
  - Should be able to take an Ethernet frame from a customer-facing port (access port), strip the CRC and preamble, and encapsulate the remaining frame using an MPLS packet.
  - An L2PE that receives an MPLS packet should be able to decide which VPLS this packet belongs to and then send it to all customer-facing ports that belong to the VPLS.
  - Maintains mapping between the learned MAC addresses and the customer's ports and mapping between learned MAC addresses and labels. This mapping constitutes the L2PE's MAC address cache.
  - Maintains a separate MAC address cache per VPLS.

- — Maintains a mapping between customer-facing ports and the different VPLS.
- — Runs a protocol between it and the PE in a client/server model. The PE has the intelligence to discover the VPLSs in the network and to inform the L2PE of right labels (or label blocks, as described earlier). The L2PE uses these labels to reach its destination.

- The PE:
  - — Needs to support the L2VPN functionality as described previously in the "L2VPN BGP Model" section. This means that a PE should be able to discover all the VPLSs in which it participates and distribute information about labels and about other L2PEs in the network.
  - — Runs a PE-to-L2PE protocol that allows the decoupling of functionality between these two devices.

Configuring the L2PE and PE

An L2PE needs to be told which VPLS it is a member of. This can be done by statically configuring a physical port or port/VLAN as part of a VPLS. In turn, for each (L2PE, VPLS) pair, the PE needs to be told the site ID of the (L2PE, VPLS). The PE also needs to be told which L2PEs it is connected to, and over which physical link and which VPLSs each L2PE participates in. This is illustrated in Figure 4-28.

Figure 4-28 *L2PE and PE Configuration*

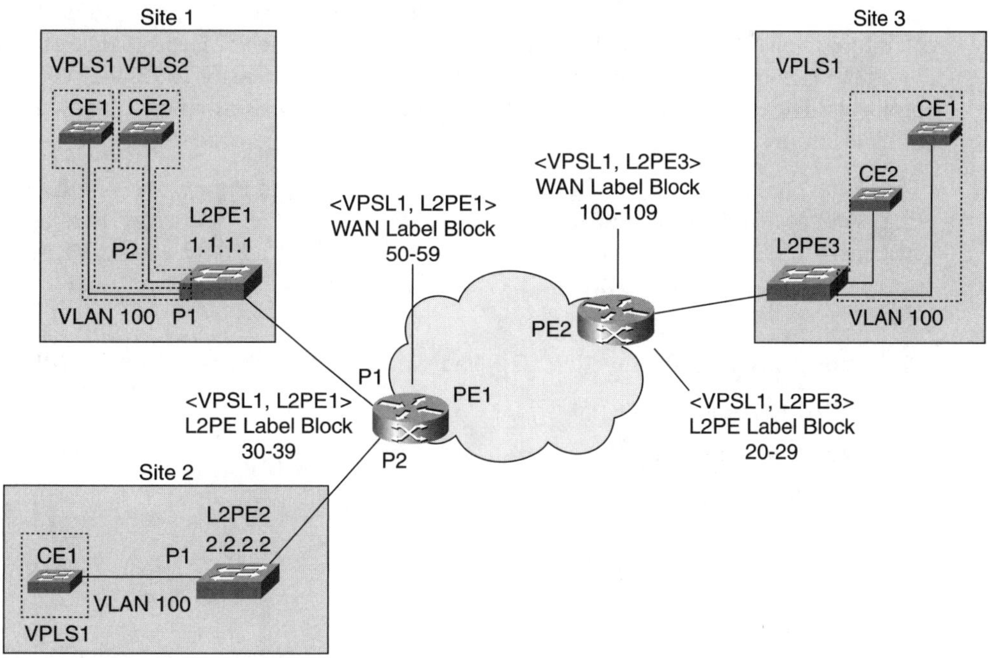

In Figure 4-28, L2PE1 is configured with the following:

- The L2PE1 ID is its router ID, 1.1.1.1
- Port 1 (P1) belongs to VPLS1
- Port 2 (P2) belongs to VPLS2
- L2PE1 is connected to PE1
- For the pair (L2PE1, VPLS1) the L2PE1 site ID is 1
- For the pair (L2PE1, VPLS2) the L2PE1 site ID is 1
- L2PE1 has a mapping between MAC addresses MAC x-MAC y with VPLS1
- L2PE1 has a mapping between MAC addresses MAC z-MAC w with VPLS2

If all information is configured on the PE, the PE can be given information pertinent to the L2PE that it can "push" into the L2PE via a certain client/server protocol. In this case, the PE needs to be configured with the following:

```
L2PE ID (router ID)
<connecting interface>
<VPLS ID> <L2PE site ID>
    <L2PE port ID, VLAN tag> <L2PE port ID, VLAN tag>
<VPLS ID> <L2PE site ID>
    <L2PE port ID, VLAN tag> <L2PE port ID, VLAN tag>
```

For each L2PE and each VPLS that the L2PE participates in, the PE is given the customer-facing port IDs and corresponding VLAN tags that belong to that VPLS. The PE then transfers all information relevant to that L2PE using the L2PE-PE protocol. The protocol that allows the information exchange between the L2PE and PE can be an extension to LDP or via other protocols. The PE transfers all information relevant to other PEs using the L2VPN BGP mechanism.

The following is a sample configuration for PE1, as shown in Figure 4-28. In this example, PE1 is connected to two different sites, 1 and 2. In site 1, L2PE1 offers service to two customers. Customer 1 on port 1 has VPLS1, which emulates a LAN between VLAN 100 across different sites (sites 1, 2, and 3). Customer 2 has VPLS2, which emulates a LAN for all VLANs.

PE1 is connected to L2PE1 and L2PE2. The following is the configuration for L2PE1 in PE1:

- L2PE1 has router ID 1.1.1.1
- Connecting interface: P1
- <VPLS 1, site ID 1>
    - <L2PE1 port 1, VLAN 100>
- <VPLS 2, site ID 1>
    - <L2PE1 port 2, all>

- For VPLS1:
  - WAN label block 50–59, label range = 10, label base = 50, block offset = 0
  - L2PE label block 30–39, label range = 10, label base = 30, block offset = 0
- For VPLS2:
  - WAN label block is x
  - L2PE label block is y

The following is the configuration for L2PE2 in PE1:

- L2PE2 has router ID 2.2.2.2
- Connecting interface: P2
- <VPLS 1, site ID 2>
  - <L2PE2 port 1, VLAN 100>
- For VPLS1:
  - WAN label is etc.
  - L2PE label block is etc.

Note that the PE1 configuration includes the indication of label blocks and label ranges. This is the same concept discussed earlier for the Frame Relay scenario; however, two sets of label blocks need to be configured for each PE. One set, called the *WAN label block,* is used to direct traffic received from L2PEs served by other PEs to the correct L2PE served by this PE. The other set of label blocks is the L2PE label block that tells the L2PEs which label to use when sending traffic to another L2PE. This creates in the network a hierarchy where the L2PEs exchange information with the connected PEs and the PEs exchange information with each other.

The site ID of an L2PE could be used as an offset from a label base to create a label. The next two sections explain how this is applied for WAN labels and the L2PE labels.

WAN Labels

For VPLS1, PE3, which is connected to L2PE3 in site 3, sends a BGP advertisement to PE1. This advertisement contains PE3's WAN label base of 100, a block offset of 0 (because only one label block is used), and the label range of 10.

For VPLS1, PE1, which is connected to L2PE1 in site 1, sends a BGP advertisement to PE3. This advertisement contains PE1's WAN label base of 50, a block offset of 0, and a label range of 10. PE1 also sends advertisements for all the <L2PE, VPLS> pairs it connects to, such as <L2PE1, VPLS2> and L2PE2 and its respective VPLS.

PE1 uses label 101 when sending packets to L2PE3. This is calculated by taking PE3's label base (100) and adding PE1's site ID (1). PE3 uses label 53 when sending packets to L2PE1. This is calculated by taking PE1's label base (50) and adding PE3's site ID (3).

L2PE Labels

Each PE also allocates a set of label blocks, called *L2PE labels,* that will be used by the L2PEs. For VPLS1, PE1 sends to L2PE1 a label base of 30, a label range of 10, and a block offset of 0. For VPLS1, PE3 sends to L2PE3 a label base of 20 and a range of 10.

L2PE1 uses label 33 when sending packets to PE1. This is calculated by taking PE1's label base (30) and adding L2PE3's site ID (3). L2PE3 uses label 21 when sending packets to PE3. This is calculated by adding PE3's label base (20) to L2PE1's site ID (1).

Following a packet in VPLS1 from L2PE1, L2PE1 takes the Ethernet frame coming from port 1, VLAN 100, and encapsulates it in an MPLS frame with label 33. PE1 receives the packet with label 33 and swaps this label with label 101, which is sent to L2PE3. PE1 encapsulates another PE-to-PE label, which directs the packet from PE1 to PE3. When the packet reaches PE3, PE3 swaps the label 103 for a label 21 and directs the packet to L2PE3. Based on this label, L2PE3 directs the packet to VPLS1.

Flooding, learning, and spanning-tree behavior at the L2PE are similar to what was previously described with the L2VPN and the LDP PW model. When a packet with an unknown destination reaches the L2PE, the L2PE identifies to which VPLS this packet belongs. It then replicates the packet over all ports in the VPLS. If the packet is received on a customer-facing port, the L2PE sends a copy out every other physical port or VLAN that participates in the VPLS, as well as to every other L2PE participating in the VPLS. If the packet is received from a PE, the packet is sent to only customer-facing ports in the MPLS, assuming that a full mesh of PEs already exists.

If an L2PE wants to flood a VPLS packet to all other L2PEs in the VPLS, the L2PE sends a copy of the packet with each label in the L2PE label ranges for that VPLS, except for the label that corresponds to the L2PE itself.

The drawback of doing the flooding at the L2PE is that the L2PE is connected to many other L2PEs in other sites and has to do quite a lot of replications. You have to weigh this against the benefits of removing the MAC learning from the PEs and keeping it in the L2PEs.

As mentioned, the protocol used for the PE-to-L2PE information exchange can be an extension of LDP. Also, there is no technical restriction on whether the tags used between the L2PE and the PE are MPLS labels. Using VLAN tags with Q-in-Q is also a possibility. The choice of one approach or the other is implementation-specific and depends on the L2PE and PE equipment capability. The upper VLAN tag sent between the L2PE and the PE is indicative of the VPLS. The PE needs to match that tag with the right WAN label to transport the packet to the remote L2PEs. It is also possible to use LDP as a universal protocol to allow the exchange of Q-in-Q tags between the PE and L2PE in the same way that MPLS labels are exchanged.

Conclusion

You have seen in this chapter how IP/MPLS can be used to scale L2 Ethernet service deployments. By keeping L2 Ethernet networks confined to the access/edge and IP/MPLS at the edge/core, service providers can leverage the simplicity of deploying Ethernet LANs with the scalability offered by IP and MPLS. L2 Ethernet services can be offered as P2P or MP2MP services. P2P can be achieved via mechanisms such as L2TPv3 or EoMPLS draft-martini. MP2MP can be achieved via VPLS.

You have seen that the flexibility VPLS offers with any-to-any connectivity is also coupled with the drawbacks of delivering Ethernet LANs in dealing with L2 loops and broadcast storms. Also with VPLS come the challenges of dealing with MAC address explosion, because PEs have to keep track of all MAC addresses advertised within the VPLS(s) the PEs belong to. Some alternatives, such as DTLS, are proposed for dealing with MAC explosion; however, different network designs and different L2PE-to-PE protocols would have to be defined and standardized.

Part II of this book, starting with Chapters 5 and 6, builds on the fact that scalable L2VPN networks are built with hybrid Ethernet and IP/MPLS networks. It also focuses on scaling the MPLS portion of the network with mechanisms such as traffic engineering via RSVP-TE and traffic protection via MPLS fast reroute. Chapters 7 and 8 move into the more advanced topic of Generalized MPLS (GMPLS). Metro networks are built with legacy TDM technology, so it is important to understand how the proliferation of MPLS in the metro will affect network provisioning on both packet and TDM networks—hence the need for a generalized control plane like GMPLS.

PART II

MPLS: Controlling Traffic over Your Optical Metro

This chapter covers the following topics:

- Advantages of Traffic Engineering
- Pre-MPLS Traffic Engineering Techniques
- MPLS and Traffic Engineering

CHAPTER 5

MPLS Traffic Engineering

You have seen in the previous chapters how metro Ethernet Layer 2 (L2) services can be deployed over an MPLS network. You also learned about the concept of pseudowires and label switched path (LSP) tunnels. The LSP tunnels are simply a means to tunnel the pseudowires from one end of the MPLS cloud to the other with the opportunity of aggregating multiple pseudowires within a single LSP tunnel. The LSP tunnels themselves can be constructed manually, or via MPLS signaling using the Label Distribution Protocol (LDP) or RSVP traffic engineering (TE). TE is an important MPLS function that gives the network operator more control over how traffic traverses the network. This chapter details the concept of TE and its use.

Advantages of Traffic Engineering

One of the main applications of MPLS is TE. A major goal of Internet TE is to facilitate efficient and reliable network operations while simultaneously optimizing network resource utilization and traffic performance. TE has become an indispensable function in many large provider networks because of the high cost of network assets and the commercial and competitive nature of the Internet.

The purpose of TE is to optimize the performance of operational networks. TE forces packets to take predetermined paths to meet network policies. In general, TE provides more efficient use of available network resources; provides control of how traffic is rerouted in the case of failure; enhances performance characteristics of the network relative to packet loss, delay, and so on; and enables value-added services, such as guaranteeing QoS and enforcing SLAs.

With metro Ethernet services, you have seen that setting bandwidth parameters on the UNI connection between the customer edge (CE) and the provider edge (PE) devices is part of the service sold to the customer. An Ethernet service with a committed information rate (CIR) of 1 Mbps should guarantee the customer that much bandwidth. It is the service provider's duty to make sure that the bandwidth promised to the customer can be allocated on the network and that the traffic adheres to the packet loss and delay parameters that are promised. TE gives the service provider more control over how traffic from multiple customers is sent over the network, enabling the service provider to make the most use of the resources available and to optimize performance.

In reference to RFC 2702, *Requirements for Traffic Engineering over MPLS,* the key performance objectives for TE can be classified as either of the following:

- Traffic-oriented
- Resource-oriented

Traffic-oriented performance objectives deal with traffic characteristics such as minimizing loss and delay to enhance traffic quality. In reference to the performance parameters defined in Chapter 3, "Metro Ethernet Services," traffic characteristics include availability, delay, jitter, and packet loss.

Resource-oriented performance objectives are mainly concerned with the optimization of resource utilization. The top priority of these objectives is to manage bandwidth resources through congestion control. Network congestion typically manifests under two scenarios:

- When network resources are insufficient or inadequate to accommodate the traffic load. An example is a spoke between a multitenant unit (MTU) device and a provider edge router at the central office (CO) that has less bandwidth than required to service all the customers of the building according to an agreed-upon SLA with the service provider.
- When traffic streams are inefficiently mapped onto available resources, causing subsets of network resources to become overutilized while others remain underutilized. An example is the existence of multiple parallel links on the backbone where some of these links are oversubscribed and are dropping traffic while others are sitting idle. This is because of how Interior Gateway Protocols (IGPs) calculate the shortest path, as explained in the next section.

Expanding capacity, or overprovisioning, alleviates the first type of congestion. Adding more or bigger network pipes is a quick and easy fix, but it comes at additional cost. Other classic congestion-control techniques, such as rate limiting, queue management, and others, can also be used to deal with insufficient network resources. These techniques are important to prevent a set of users or traffic types from consuming the whole bandwidth and starving other users on the network.

This chapter mainly addresses the second type of congestion problems—those resulting from inefficient resource allocation. You can usually address these congestion problems through TE. In general, you can reduce congestion resulting from inefficient resource allocation by adopting load-balancing policies. The objective of such strategies is to minimize maximum congestion, or alternatively to minimize maximum resource utilization, through efficient resource allocation. When congestion is minimized through efficient resource allocation, packet loss decreases, transit delay decreases, and aggregate throughput increases. This significantly enhances the end users' perception of network service quality.

As you have noticed, this chapter so far hasn't mentioned MPLS, because TE by itself is universal and a well-understood problem. The use of MPLS for TE is one of the methods for dealing with resource optimization, and the industry has begun adopting MPLS techniques only after going through many alternatives to solve the TE problem. The next section discusses some of the pre-MPLS TE techniques.

Pre-MPLS Traffic Engineering Techniques

Pre-MPLS TE techniques involved multiple mechanisms:

- Altering IGP routing metrics
- Equal-cost multipath
- Policy-based routing
- Offline design of virtual circuit overlays

Altering IGP Routing Metrics

IGPs have many limitations when used to achieve traffic engineering. IGPs rely on metrics that do not reflect actual network resources and constraints. IGPs based on Shortest Path First (SPF) algorithms contribute significantly to congestion problems in IP networks. SPF algorithms generally optimize based on a simple additive metric. These protocols are topology-driven, so real-time bandwidth availability and traffic characteristics are not factors considered in routing decisions. As such, congestion occurs when the shortest paths of multiple streams converge over one link that becomes overutilized while other existing links are underutilized, as shown in Figure 5-1.

Figure 5-1 *IGP Shortest Path First Congestion*

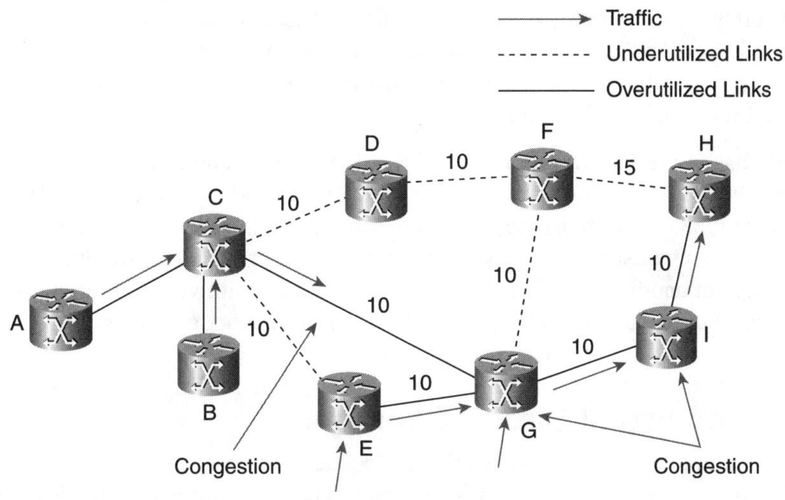

In Figure 5-1, based on the indicated link metric, an OSPF routing algorithm allows traffic coming from routers A and B, destined for router H, to use path C-G-I-H. Traffic from routers E and G, destined for H, uses path G-I-H. As you can see, multiple streams of traffic have converged on the same links or routers, causing congestion on link G-I, for example, while other links in the network remain underutilized.

Altering IGP metrics could cause traffic to shift between links. Changing the metric of link F-H from 15 to 10 or 5 could cause the traffic to start taking links C-D-F-H or C-G-F-H. Link manipulation for the purposes of TE works for quick-fix solutions but is mainly a trial-and-error process and does not scale as an operational model. Adjusting the link metrics might fix one problem but create other problems in the network.

Equal-Cost Multipath

Equal-cost multipath is a mechanism that allows routers to distribute traffic between equal-cost links to efficiently use the network resources and avoid the problem of link oversubscription and undersubscription. If, for example, a router calculates the shortest path based on link metrics and determines multiple equal paths exist to the same destination, the router can use load-balancing techniques to spread the traffic flows over the equal-cost links. Referring to Figure 5-1, if the metric of link F-H is changed to 10 instead of 15, the paths C-D-F-H and C-G-I-H would have the same metric, 30 (10 + 10 + 10). Traffic from routers A and B, destined for H, could be load balanced across the two equal-cost paths.

Policy-Based Routing

Policy-based routing is another mechanism that can be used for TE. It allows routers to dictate the traffic trajectory. That is, they pick the router output interface on which to route traffic based on a policy—for example, based on the source IP address rather than the destination IP address. With this type of TE, you can dictate that traffic coming from a certain customer or provider goes one way, while traffic from other customers or providers goes the other way, irrespective of what its actual destination is.

Policy-based mechanisms can be used to allow more granularity in identifying the source traffic. For example, the traffic can be identified based on source IP address, router port numbers, QoS, application type, and so on. Although this type of TE is useful, it has its drawbacks. First, it acts against the nature of routing, which is primarily destination-based. Second, it becomes yet another form of intelligent static routing with vulnerability to traffic loops and to the lack of dynamic rerouting in case of failure of network elements.

Offline Design of Virtual Circuit Overlays

A popular approach to circumvent the inadequacies of current IGPs is to use an overlay model, such as IP over ATM or IP over Frame Relay. The overlay model extends the design space by enabling arbitrary virtual topologies to be provisioned on top of the network's physical topology. The virtual topology is constructed from virtual circuits (VCs) that appear as physical links to the IGP routing protocols. Overlay techniques can range from simple permanent virtual circuit (PVC) provisioning between routed edge networks to more fancy mechanisms that include constraint-based routing at the VC level with support of configurable explicit VC paths, traffic shaping and policing, survivability of VCs, and so on.

Figure 5-2 shows edge routers that are connected to each other via an overlay model on top of an ATM network. For the IGPs, the VCs appear as direct physical links between the routers. Traffic can be engineered between the routed edges and is agnostic to the L2 switched network in the middle of the cloud. This type of TE has several benefits: It enables you to achieve full traffic control, measure link statistics, divert traffic based on link utilization, apply load-balancing techniques, and so on. It also has several obvious drawbacks: It creates multiple independent control planes—IP and ATM—that act independently of one another, a full mesh between the routers, an IGP neighbor explosion (each router has all other routers as neighbors and has to exchange routing updates with them), and finally a network management challenge constituting multiple control layers.

Figure 5-2 *IGP TE Via Virtual Circuit Overlays*

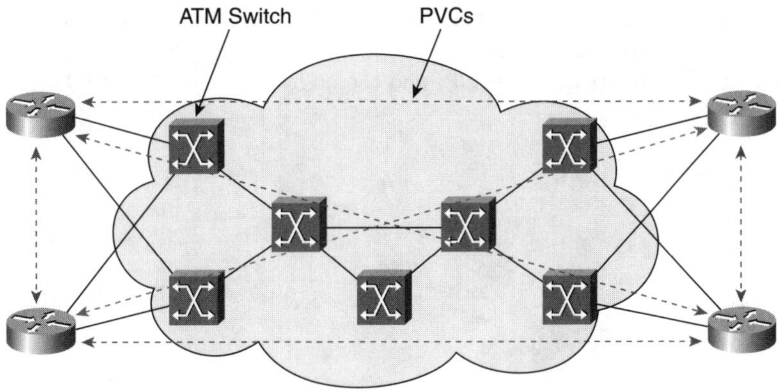

MPLS and Traffic Engineering

MPLS is strategically significant for TE because it can potentially provide most of the functionality available from the overlay model (described in the preceding section), with much better integration with IP. MPLS for TE is attractive because it enables you to do the following:

- Manually or dynamically build explicit LSPs
- Efficiently manage LSPs
- Define traffic trunks and map them to LSPs
- Associate a set of attributes with traffic trunks to change their characteristics
- Associate a set of attributes with resources that constrain the placement of LSPs and traffic trunks mapped to those LSPs
- Aggregate and deaggregate traffic (whereas IP routing only allows aggregation)

- Easily incorporate a constraint-based routing framework with MPLS
- Deliver good traffic implementation with less overhead than pre-MPLS techniques
- Define backup paths with fast failover

Before delving into more details about TE, it helps to explain the terminology of trunks versus LSPs, because the two are often confused with one another.

Traffic Trunks Versus LSPs

Traffic trunks are not LSPs. The definition of traffic trunks as indicated in RFC 2430 follows: "A traffic trunk is an aggregation of traffic flows of the same class which are placed inside an LSP." Examples of flow classes can be similar to Diffserv. Traffic trunks are also routable objects, similar to VCs for ATM. A traffic trunk can be mapped to a set of LSPs and can be moved from one LSP to another.

An LSP, on the other hand, is a specification of the path through which the traffic traverses. The LSP is constructed through label swapping between ingress to egress to switch the traffic to its destination. Trunks traverse LSPs and can be routed from one LSP to another. This is illustrated in Figure 5-3.

Figure 5-3 *Trunks and LSPs*

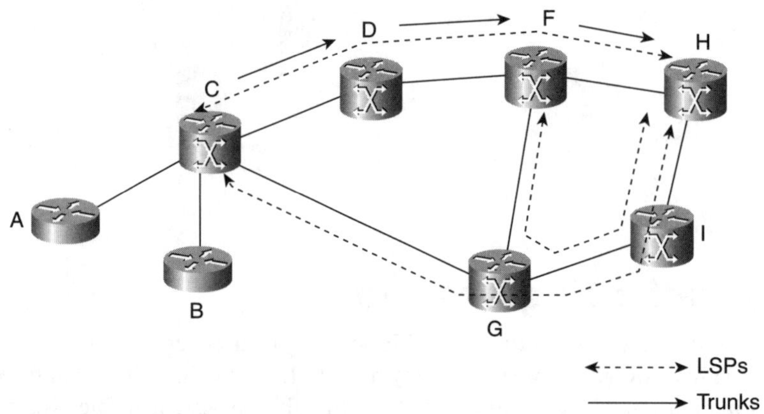

Figure 5-3 shows two LSPs between routers C and H, LSP C-D-F-H and LSP C-G-I-H. Another LSP exists between routers F and H, LSP F-G-I-H. A set of traffic flows belonging to the same class, coming from router A and destined for destinations beyond router H, could be mapped to either LSP C-D-F-H or LSP C-G-I-H. This aggregated traffic flow is the traffic trunk. The same traffic trunk can be routed over LSP F-G-I-H if some trunk attributes, such as resiliency or bandwidth, are being enforced.

Capabilities of Traffic Engineering over MPLS

The functional capabilities required to support TE over MPLS in large networks involve the following:

- A set of attributes that affect the behavior and characteristics of traffic trunks
- A set of attributes that are associated with resources and that constrain the placement of traffic trunks over LSPs
- A constraint-based routing framework that is used to select paths subject to constraints imposed by traffic trunk attributes and available resources

The attributes associated with traffic trunks and resources, as well as parameters associated with routing, represent a set of variables that can be used to engineer the network. These attributes can be set either manually or through automated means. The next section discusses traffic trunk operation and attributes.

Traffic Trunk Operation and Attributes

Traffic trunks are by definition unidirectional, but it is possible to instantiate two trunks in opposite directions with the same endpoints. The set of traffic trunks, one called *forward trunk* and the other called *backward trunk,* form a logical bidirectional traffic trunk. The bidirectional traffic trunks can be topologically symmetrical or asymmetrical. A bidirectional traffic trunk is symmetrical if opposite trunks take the same physical path, and it is asymmetrical if opposite trunks take different physical paths.

The basic operations that you can perform on a trunk include establishing a trunk; activating, deactivating, and destroying a trunk; modifying a trunk's attributes; and causing a trunk to reroute from its original path via manual or dynamic configuration. You can also police the traffic to comply with a certain SLA and shape and smooth the traffic before it enters the network.

As described in RFC 2702, the following are the basic attributes of traffic trunks that are particularly significant for TE:

- Traffic parameter attributes
- Generic path selection and maintenance attributes
- Priority attribute
- Preemption attribute
- Resilience attribute
- Policing attribute
- Resource attributes

Traffic Parameter Attributes

Traffic parameter attributes indicate the resource requirements of a traffic trunk that are useful for resource allocation and congestion avoidance. Such attributes include peak rates, average rates, permissible burst size, and so on. Chapter 3 describes the applicable parameters, such as committed information rate (CIR), peak information rate (PIR), and so on.

Generic Path Selection and Maintenance Attributes

Generic path selection and maintenance attributes define how paths are selected, such as via underlying network protocols, via manual means, or via signaling. If no restrictions exist on how a traffic trunk is established, IGPs can be used to select a path. If restrictions exist, constraint-based routing signaling, such as RSVP-TE, should be used.

Chapter 4, "Hybrid L2 and L3 IP/MPLS Networks," describes how a metro provider carries L2 services over an MPLS cloud via the use of pseudowires (VC-LSPs) carried in LSP tunnels. If the LSP tunnels are not traffic-engineered, the traffic on the MPLS cloud follows the path dictated by the IGP. If multiple IGP paths collide, traffic congestion could occur. Setting resource requirements coupled with TE alleviates this problem.

Priority Attribute

The priority attribute defines the relative importance of traffic trunks. Priorities determine which paths should be used versus other paths at connection establishment and under fault scenarios. A metro operator could deliver Internet service as well as IP storage backhaul over different pseudowires. The IP storage traffic could be carried over a separate LSP tunnel and given a high priority to be rerouted first in case of failure.

Preemption Attribute

The preemption attribute determines whether a traffic trunk can preempt another traffic trunk from a given path. Preemption can be used to ensure that high-priority traffic can always be routed in favor of lower-priority traffic that can be preempted. Service providers can use this attribute to offer varying levels of service. A service that has preemption could be priced at a higher rate than a regular service.

Resilience Attribute

The resilience attribute determines the behavior of a traffic trunk when fault conditions occur along the path through which the traffic trunk traverses. The resiliency attribute indicates whether to reroute or leave the traffic trunk as is under a failure condition. More extended resilience attributes could specify detailed actions to be taken under failure, such as the use of alternate paths, and specify the rules that govern the selection of these paths.

Policing Attribute

The policing attribute determines the actions that should be taken by the underlying protocols when a traffic trunk exceeds its contract as specified in the traffic parameters. Policing is usually done on the input of the network, and it indicates whether traffic that does not conform to a certain SLA should be passed, rate limited, dropped, or marked for further action.

Resource Attributes

Resource attributes constrain the placement of traffic trunks. An example of resource attributes is the maximum allocation multiplier. This attribute applies to bandwidth that can be oversubscribed or undersubscribed. This attribute is comparable to the subscription and booking factors in ATM and Frame Relay. A resource is overallocated or overbooked if the sum of traffic from all traffic trunks using that resource exceeds the resource capacity. Overbooking is a typical mechanism used by service providers to take advantage of the traffic's statistical multiplexing and the fact that peak demand periods for different traffic trunks do not coincide in time.

Another example of resource attributes is the resource class attribute, which attempts to give a "class" to a set of resources. Resource class attributes can be viewed as "colors" assigned to resources such that resources with the same "color" conceptually belong to the same class. The resource class attribute can be used to implement many policies with regard to both traffic- and resource-oriented performance optimization. Resource class attributes can be used, for example, to implement generalized inclusion and exclusion to restrict the placement of traffic trunks to a specific subset of resources.

Constraint-Based Routing

Constraint-based routing assists in performance optimization of operational networks by finding a traffic path that meets certain constraints. Constraint-based routing is a demand-driven, reservation-aware routing mechanism that coexists with hop-by-hop IGP routing.

Constraints are none other than the attributes that were previously discussed: Trunk attributes such as path selection attributes, policing, preemption, and so on, coupled with resource attributes and some link-state parameter, would affect the characteristics and behavior of the traffic trunk.

A constraint-based routing framework can greatly reduce the level of manual configuration and intervention required to set TE policies. In practice, an operator specifies the endpoints of a traffic trunk and assigns a set of attributes to the trunk. The constraint-based routing framework is then expected to find a feasible path to satisfy the expectations. If necessary, the operator or a TE support system can then use administratively configured explicit routes to perform fine-grained optimization.

Figures 5-4a and 5-4b show two different types of routing applied to the same scenario. In Figure 5-4a, simple IGP routing is applied, and the shortest path is calculated based on the IGP metrics. A traffic trunk coming from router A is mapped to path (LSP) C-E-G-I-H. In Figure 5-4b, constraints are imposed on the routing construct. The constraint is *not* to use any path that has available bandwidth less than 250 Mbps. As such, the two links E-G and G-I have been removed, or pruned, from the selection algorithm, and the traffic trunk coming from router A has been mapped to path C-D-G-F-H.

Figure 5-4a *Aggregating Trunks into Tunnels*

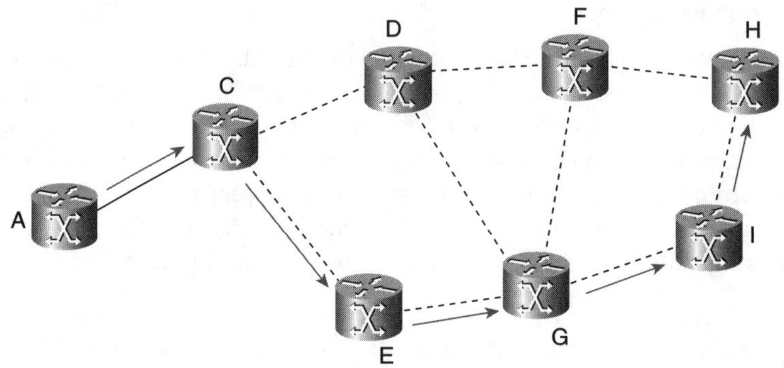

Figure 5-4b *Constraint-Based Routing*

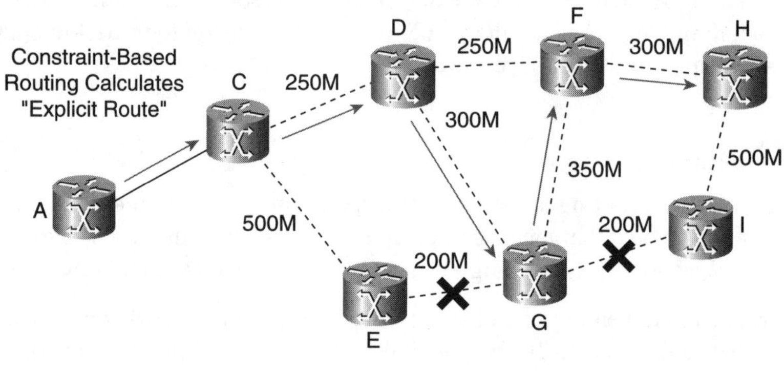

Conclusion

This chapter has discussed the different parameters used for TE. Some of the concepts, such as traffic parameter attributes and policing attributes, were discussed in the context of metro deployments in Chapter 3.

The next steps for TE entail a mechanism for exchanging the traffic attributes and parameters in the network for each router to build a TE database. This database gives the routers visibility to all the network resources and attributes that can be used as input into a Constrained Shortest Path First (CSPF) algorithm. CSPF determines the path in the network based on different constraints and attributes. Finally, a signaling protocol such as RSVP-TE is used to signal the LSP in the network based on the path determined by the CSPF. The next chapters explain the concepts behind RSVP-TE to familiarize you with how Label Switched Path are signaled across a packet network. The book extends this concept further in Chapters 7 and 8 to discuss how MPLS signaling and routing can be extended as well to cover nonpacket networks, such as the case of an optical metro.

This chapter covers the following topics:

- Understanding RSVP-TE
- Understanding MPLS Fast Reroute

CHAPTER 6

RSVP for Traffic Engineering and Fast Reroute

Traffic engineering allows the service provider to manipulate the traffic trajectory to map traffic demand to network resources. You have seen in Chapter 5, "MPLS Traffic Engineering," that traffic engineering can be achieved by manipulating Interior Gateway Protocol (IGP) metrics or, better yet, by using a signaling protocol such as RSVP-TE. RSVP-TE offers the ability to move trunks away from the path selected by the ISP's IGP and onto a different path. This allows a network operator to route traffic around known points of congestion in the network, thereby making more efficient use of the available bandwidth. It also allows trunks to be routed across engineered paths that provide guaranteed service levels, enabling the sale of classes of service.

In metro networks, traffic engineering goes hand in hand with traffic path restoration upon failures. The behavior of a network upon failure depends on what layer the restoration methods are applied to. SONET/SDH networks, for example, can achieve restoration at Layer 1, meaning that if part of a SONET/SDH ring fails, there is always a backup TDM circuit provisioned on another fiber (unidirectional path switched ring, UPSR) or another pair of fibers (bidirectional line switched ring, BLSR). With Resilient Packet Ring (RPR), the ring is always fully utilized and a failure will cause the ring to wrap, allowing the rest of the ring to remain functional.

Restoration can also be done at Layer 2. Spanning Tree Protocol (STP) and Rapid Spanning Tree Protocol (RSTP) (802.1w) are typical methods that allow the network to converge after failure. Layer 3 methods can also be used. Routing protocols such as Open Shortest Path First (OSPF) and Intermediate System-to-Intermediate System (IS-IS) are capable of computing multiple paths to the same destination. If the main path fails, the protocols converge to an alternate path. Mechanisms like equal-cost multipaths can also be used to allow faster convergence by having parallel active paths to the same destination.

In the use of traffic engineering and traffic restoration methods, operators look for the following:

- The ability to maintain customer SLAs in case of a network failure.
- The ability to achieve the most efficient use of network resources, in a way that provides good QoS that meets an SLA with their customers.
- The ability to restore failure within a timeline that does not violate any SLAs they established with their customers.

Because MPLS plays a big role in delivering and scaling services in the metro, it is important to understand how it can be used to achieve TE and protection via the use of RSVP-TE. In this chapter, you see how MPLS, through the use of RSVP-TE, can be used to establish backup paths in the case of failure.

Understanding RSVP-TE

MPLS TE may be used to divert traffic over an explicit route. The specification of the explicit route is done by enumerating an explicit list of the routers in the path. Given this list, TE trunks can be constructed in a variety of ways. For example, a trunk could be manually configured along the explicit path. This involves configuring each router along the path with state information for forwarding the particular MPLS label.

Alternately, a signaling protocol such RSVP-TE can be used with an EXPLICIT_ROUTE object (ERO) so that the first router in the path can establish the trunk. The ERO is basically a list of router IP addresses.

NOTE Constraint-based routing LDP (CR-LDP) is another signaling protocol that can be used to build traffic-engineered paths. However, the use of RSVP-TE is more widely deployed and as such will be discussed in this book.

Originally, RSVP (defined in RFC 2205, *Resource ReSerVation Protocol—Version 1 Functional Specification*) was designed as a protocol to deliver QoS in the network by allowing routers to establish resource reservation state for individual flows originated between hosts (computers). This model has not taken off with network operators because of scalability issues in maintaining the per-flow state between pairs of hosts in each router along the IGP path. The RSVP implementation is illustrated in Figure 6-1.

Figure 6-1 *Original RSVP Implementation*

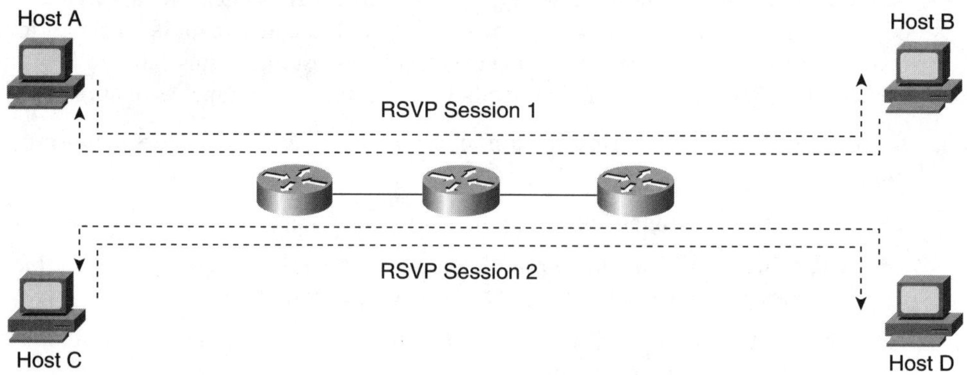

Figure 6-1 illustrates two RSVP sessions between hosts A and B and hosts C and D. The routers in the path would have to maintain state information for these sessions to allocate certain bandwidth to the individual flows. With a large number of hosts (millions) in a public network, this model has not proven to be efficient and hence has not been adopted in the public Internet.

In the late 1990s, RSVP was extended to support the creation of MPLS label switched paths (LSPs). The extended RSVP implementations introduced a lot of changes to the traditional RSVP, to support scalability issues and TE. In particular, RSVP sessions take place between ingress and egress label switch routers (LSRs) rather than individual hosts. The aggregated traffic flow, called a *traffic trunk,* is then mapped to LSPs, also called *LSP tunnels*. The RSVP-TE implementation is shown in Figure 6-2.

Figure 6-2 *RSVP-TE Implementation*

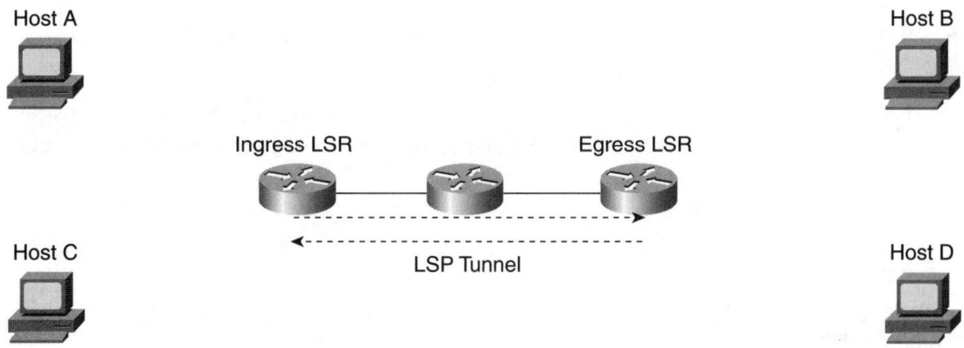

The extensions of RSVP to support MPLS and TE can accomplish the following:

- **Establish a forwarding path**—RSVP can be used to establish LSPs by exchanging label information. This mechanism is similar to the Label Distribution Protocol (LDP).
- **Establish an explicit path**—RSVP-TE is used to establish an LSP along an explicit route according to specific constraints. LSPs can be rerouted upon failure. (Fast reroute is discussed later, in the section "Understanding MPLS Fast Reroute.")
- **Resource reservation**—The existing RSVP procedures for resource reservation can be applied on aggregated flows or traffic trunks. This model scales because it is done on trunks rather than flows and is done between pairs of routers rather than pairs of hosts, as was originally intended for RSVP.

The reason IETF chose to extend RSVP to support MPLS and TE has to do with the fact that RSVP was originally designed for resource reservation in the Internet, a concept that is closely tied to TE, so it makes sense to extend the protocol rather than create a new one. RSVP also can carry opaque objects such as fields that can be delivered to routers, which makes it easy to define new objects for different purposes. The purpose of some of these objects is to carry labels for the label distribution function, whereas the purpose of others is to create explicit routes.

The following sections describe how RSVP tunnels are created and the mechanisms that are used to exchange MPLS labels and reserve bandwidth:

- RSVP LSP Tunnels
- Label Binding and LSP Tunnel Establishment Via RSVP
- Reservation Styles
- Details of the PATH Message
- Details of the Reservation Message

RSVP LSP Tunnels

Service providers create LSP tunnels to aggregate traffic belonging to the same forwarding equivalency class. You have seen in Chapter 4, "Hybrid L2 and L3 in IP/MPLS Networks," that multiple Virtual Private LAN Services (VPLSs) can be carried over a single LSP tunnel across the network.

LSPs are called LSP tunnels because the traffic going through an LSP tunnel is opaque to the intermediate LSRs between the ingress and egress LSRs. Figure 6-3 shows the establishment of an LSP tunnel between an ingress LSR and an egress LSR that is peering with multiple providers. Notice how the LSP tunnel is formed using two unidirectional tunnels in both directions.

Figure 6-3 *LSP Tunnel Between Ingress and Egress LSRs*

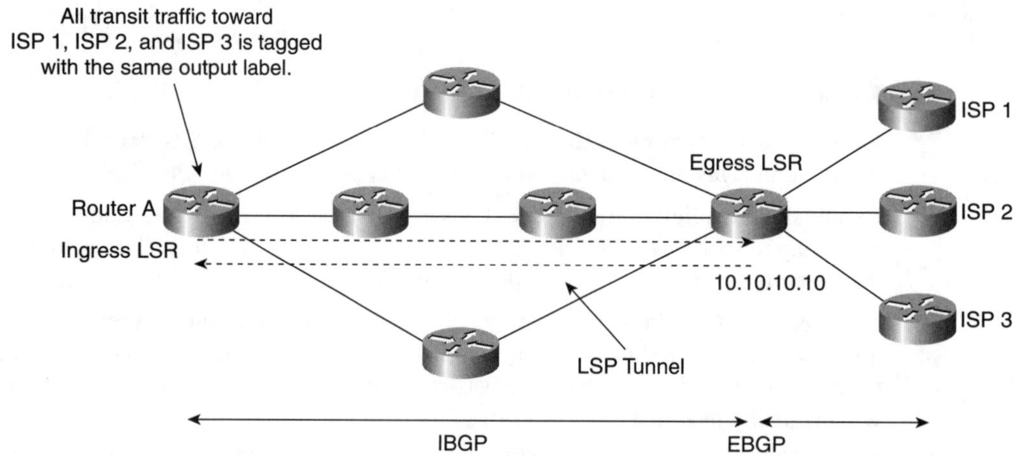

In Figure 6-3, traffic coming into router A and transiting the service provider's network toward other service providers' networks, such as ISP 1, ISP 2, and ISP 3, can all be grouped in the same forwarding equivalency class. This class is defined by all traffic destined for the exit router with IP address 10.10.10.10. In this case, all traffic toward 10.10.10.10 is tagged with the same outbound label at router A. This maps all transit traffic toward the same LSP tunnel.

The exit point for a given external route (10.10.10.10) is normally learned via the Internal Border Gateway Protocol (IBGP). After the traffic reaches the exit router, it is sent to the correct ISP, depending on the final external route.

Label Binding and LSP Tunnel Establishment Via RSVP

RFC 3209, *RSVP-TE: Extensions to RSVP for LSP Tunnels,* defines the capabilities of extended RSVP. Regarding the operation of LSP tunnels, extended RSVP enables you to do the following:

- Perform downstream-on-demand label allocation, distribution, and binding.
- Observe the actual route traversed by an established LSP tunnel.
- Identify and diagnose LSP tunnels.
- Establish LSP tunnels with or without QoS requirements.
- Dynamically reroute an established LSP tunnel.
- Preempt an established LSP tunnel under administrative policy control.

To establish an LSP tunnel, the ingress LSR sends a PATH message to the egress LSR, which in turn replies with a reservation message (RESV). Upon completion of the handshake, an LSP tunnel is established. The PATH message indicates the PATH that the LSP should take, and the RESV message attempts to establish a bandwidth reservation following the opposite direction of the PATH message. PATH and RESV messages are explained in detail in the sections "Details of the PATH Message" and "Details of the RESV Message," respectively.

RSVP-TE has defined new objects in support of creating LSP tunnels. These new objects, called LSP_TUNNEL_IPv4 and ISP_Tunnel_IPv6, help, among other things, identify LSP tunnels. The SESSION object, for instance, carries a tunnel ID, while the SENDER_TEMPLATE and FILTER_SPEC objects uniquely identify an LSP tunnel.

The following is the sequence of events needed to establish an LSP tunnel:

1. The first MPLS node on the path—that is, the ingress LSR (sender)—creates an RSVP PATH message with a session type of LSP_TUNNEL_IPv4 or LSP_TUNNEL_IPv6 and inserts a LABEL_REQUEST object into the PATH message.
2. The LABEL_REQUEST object indicates that a label binding for this path is requested and also indicates the network layer protocol that is to be carried over this path.

   In addition to the LABEL_REQUEST object, the PATH message can carry a number of optional objects:

   - **EXPLICIT_ROUTE object (ERO)**—Specifies a predetermined path between the ingress and egress LSRs. When the ERO object is present, the PATH message is sent toward the first node indicated by the ERO, independent of the IGP shortest path.

- **RECORD_ROUTE object (RRO)**—Used to record information about the actual route taken by the LSP. This information can be relayed back to the sender node. The sender node can also use this object to request notification from the network concerning changes in the routing path.
- **SESSION_ATTRIBUTE object**—Can be added to PATH messages to help in session identification and diagnostics. Additional control information, such as setup and hold priorities and local protection, is also included in this object.

3 The label allocation with RSVP is done using the downstream-on-demand label assignment mechanism.

4 The RESV message is sent back upstream toward the sender, following the path created by the PATH message, in reverse order.

5 Each node that receives an RESV message containing a LABEL object uses that label for outgoing traffic associated with this LSP tunnel.

6 When the RESV message arrives at the ingress LSR, the LSP tunnel is established.

This process is illustrated in Figure 6-4.

Figure 6-4 *Establishing an LSP Tunnel*

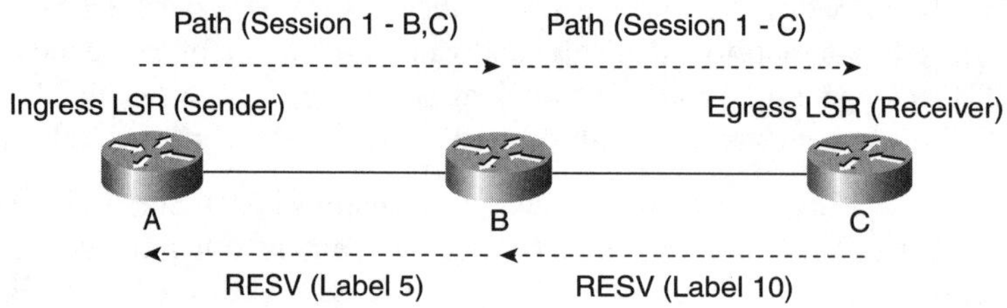

In Figure 6-4, ingress LSR A sends a PATH message toward LSR C with a session type object and an ERO. The ERO contains the explicit route that the PATH message needs to take. The ERO in this case is the set {B,C}, which dictates the path to be taken via LSR B, then LSR C.

In turn, LSR B propagates the PATH message toward LSR C according to the ERO. When LSR C receives the PATH message, it sends an RESV message that takes the reverse PATH indicated in the ERO toward LSR A. LSR C includes an inbound label of 10. Label 10 is used as an outbound label in LSR B. LSR B sends an RESV message toward LSR A with an inbound label of 5. Label 5 is used as an outbound label by LSR A. An LSP tunnel is formed between LSRs A and C. All traffic that is mapped to this LSP tunnel is tagged with label 5 at LSR A.

Reservation Styles

The existing RSVP procedures for resource reservation can be applied on aggregated flows or traffic trunks. This model scales because it is done on trunks rather than flows and between pairs of routers rather than pairs of hosts, as was originally intended for RSVP. The receiver node can select from among a set of possible reservation styles for each session, and each RSVP session must have a particular style. Senders have no influence on the choice of reservation style. The receiver can choose different reservation styles for different LSPs. Bandwidth reservation is not mandatory for the operation of RSVP-TE. It is up to the service provider to engineer the networks as necessary to meet the SLAs.

The following sections discuss the different reservation styles listed here and their advantages and disadvantages:

- Fixed Filter (FF)
- Shared Explicit (SE)
- Wildcard Filter (WF)

Fixed Filter Reservation Style

The FF reservation style creates a distinct reservation for traffic from each sender. This style is normally used for applications whose traffic from each sender is independent of other senders. The total amount of reserved bandwidth on a link for sessions using FF is the sum of the reservations for the individual senders. Because each sender has its own reservation, a unique label is assigned to each sender. This can result in a point-to-point LSP between every sender/receiver pair. An example of such an application is a one-on-one videoconferencing session. Bandwidth reservations between different pairs of senders and receivers are independent of each other.

In Figure 6-5, ingress LSRs A and B create distinct FF-style reservations toward LSR D. The total amount of bandwidth reserved on link C-D is equal to the sum of reservations requested by A and B. Notice also that LSR D has assigned different labels in the RESV messages toward A and B. Label 10 is assigned for sender A, and label 20 is assigned for sender B. This creates two distinct point-to-point LSPs—one between A and D and the other between B and D.

Figure 6-5 *Fixed Filter Reservation Style*

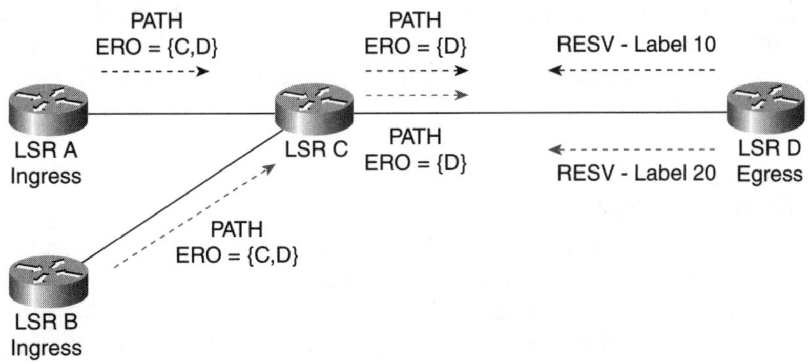

Shared Explicit Reservation Style

The SE reservation style allows a receiver to explicitly select a reservation for a group of senders—rather than one reservation per sender, as in the FF style. Only a single reservation is shared between all senders listed in the particular group.

SE style reservations can be provided using one or more multipoint-to-point LSPs per sender. Multipoint-to-point LSPs may be used when PATH messages do not carry the ERO, or when PATH messages have identical EROs. In either of these cases, a common label may be assigned.

PATH messages from different senders can each carry their own ERO, and the paths taken by the senders can converge and diverge at any point in the network topology. When PATH messages have differing EROs, separate LSPs for each ERO must be established. Figure 6-6 explains the SE style even further.

Figure 6-6 *Shared Explicit Reservation Style*

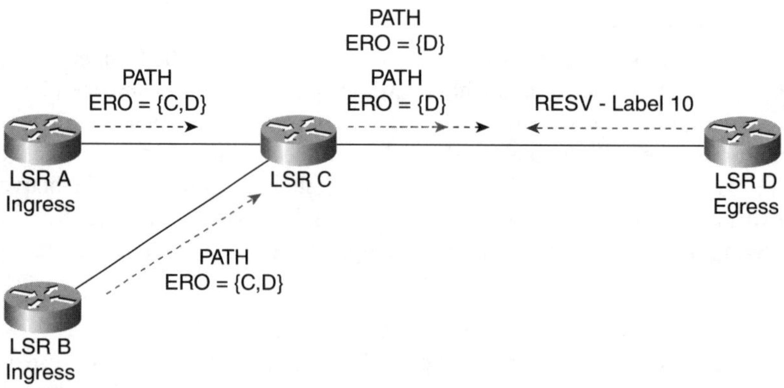

In Figure 6-6, LSRs A and B are using the SE style to establish a session with LSR D. The reservation for link C-D is shared between A and B. In this example, both PATH messages coming from A and B have the same ERO and are converging on node C. Notice that D has allocated a single label 10 in its RESV message, hence creating the multipoint-to-point LSP. An example of such an application is a videoconferencing session between multiple branch offices in Europe and the main office in the United States. The bandwidth reserved on the international link is set for a certain amount, and the number of remote branch offices is set in a way that the total amount of bandwidth used by the branch offices does not exceed the total reserved bandwidth.

Wildcard Filter Reservation Style

A third reservation style that is defined by RSVP is the WF reservation style. Unlike the SE style, where the receiver indicates the specific list of senders that are to share a reservation, with the WF reservation style, a single shared reservation is used for all senders to a session. The total reservation on a link remains the same regardless of the number of senders.

This style is useful for applications in which not all senders send traffic at the same time. If, however, all senders send simultaneously, there is no means of getting the proper reservations made. This restricts the applicability of WF for TE purposes.

Furthermore, because of the merging rules of WF, EROs cannot be used with WF reservations. This is another reason that prevents the use of the WF style for traffic engineering.

Details of the PATH Message

The PATH message can include several different RSVP objects, including the following:

- LABEL_REQUEST
- EXPLICIT_ROUTE
- RECORD_ROUTE
- SESSION_ATTRIBUTE
- FLOW_SPEC
- SENDER_TEMPLATE
- SESSION

Figure 6-7 shows the format of the PATH message.

Figure 6-7 *RSVP-TE PATH Message*

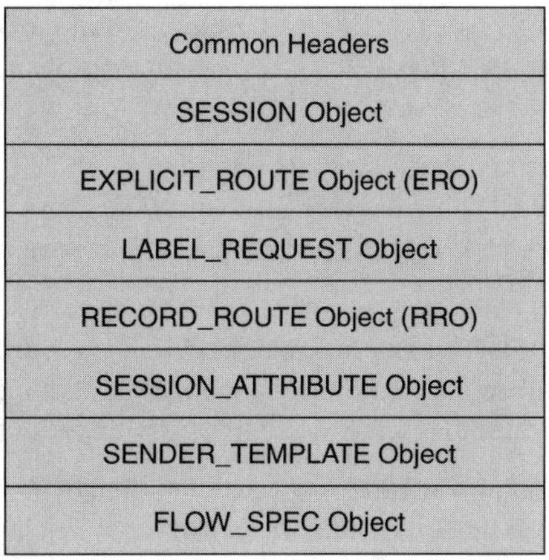

The following sections describe each object in more detail.

LABEL_REQUEST Object

The LABEL_REQUEST object is used to establish label binding for a certain path. It also indicates the network layer protocol that is to be carried over this path. The reason for this is that the network layer protocol sent down an LSP does not necessarily have to be IP and cannot be deduced from the L2 header, which only identifies the higher-layer protocol as MPLS. The LABEL_REQUEST object has three possible C_Types (Class_Types):

- **Type 1, label request without label range**—This is a request for a regular 32-bit MPLS label that sits in the shim layer between the data link and network layer headers.
- **Type 2, label request with an ATM label range**—This request specifies the minimum and maximum virtual path identifier (VPI) and virtual connection identifier (VCI) values that are supported on the originating switch. This is used when the MPLS label is carried in an ATM header.
- **Type 3, label request with Frame Relay label range**—This request specifies the minimum and maximum data-link connection identifier (DLCI) values that are supported on the originating switch. This is used when the MPLS label is carried in a Frame Relay header.

When the PATH message reaches an LSR, the LABEL_REQUEST object gets stored in the path state block for further use by refresh messages. When the PATH message reaches the receiver, the presence of a LABEL_REQUEST object triggers the receiver to allocate a label and to place the label in the LABEL object for the corresponding RESV message. If a label range is specified, the label must be allocated from that range. Error messages might occur in cases where the receiver cannot assign a label, cannot recognize the protocol ID, or cannot recognize the LABEL_REQUEST object.

EXPLICIT_ROUTE Object

The EXPLICIT_ROUTE object (ERO) is used to specify an explicit path across the network independent of the path specified by the IGP. The contents of an ERO are a series of variable-length data items called *subobjects.*

A subobject is an abstract node that can be either a single node or a group of nodes such as an autonomous system. This means that the explicit path can cross multiple autonomous systems, and the hops within each autonomous system are opaque (hidden) from the ingress LSR for that path.

The subobject contains a 1-bit Loose Route field (L). If set to 1, this field indicates that the subobject is a loose hop in the explicit path, and if set to 0, it indicates that the subobject is a strict hop. A strict hop indicates that this hop is physically adjacent to the previous node in the path.

The subobject also contains a Type field, which indicates the types of the content subobjects. Some defined values of the Type field are as follows:

- **1: IPv4 Prefix**—Identifies an abstract node with a set of IP prefixes that lie within this IPv4 prefix. A prefix of length 32 is a single node (for example, a router's IP loopback address).
- **2: IPv6 Prefix**—Identifies an abstract node with a set of IP prefixes that lie within this IPv6 prefix. A prefix of length 128 is a single node (for example, a router's IP loopback address).
- **32: Autonomous System number**—Identifies an abstract node consisting of the set of nodes belonging to the autonomous system.

Figures 6-8 and 6-9 illustrate two scenarios in which an explicit path is being established using strict and loose subobjects, respectively, of the Type IPv4 prefix and with a subobject length of 32.

In Figure 6-8, ingress LSR A sends a PATH message toward LSR D with an ERO that indicates a strict hop across routers B (192.213.1.1), C (192.213.2.1), and D (192.213.3.1). When B receives the PATH message, it propagates it toward C, and C propagates the message toward D. In turn, D sends a RESV message to A along the same path, and the label binding takes place. The ERO itself is modified at each hop. Each node in the ERO list removes itself from the ERO as the PATH message is forwarded.

Figure 6-8 *Explicit Route, Strict Hops*

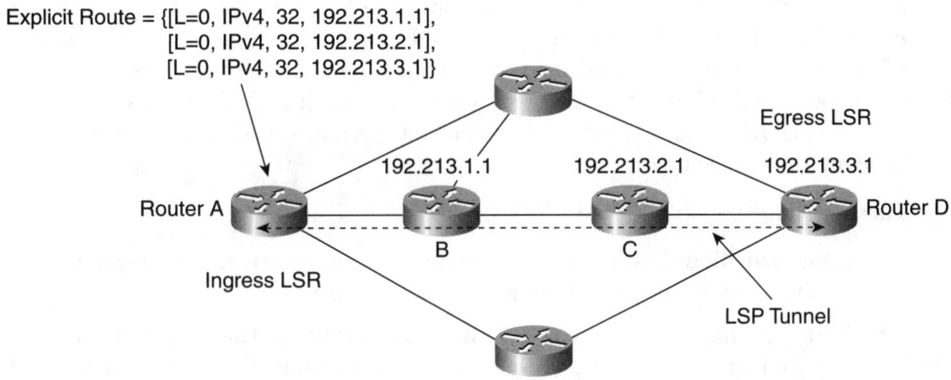

In Figure 6-9, ingress LSR A sends a PATH message toward LSR D with an ERO that indicates a strict hop toward B. From router B, a loose hop is used. When router B receives the PATH message, it would send the PATH message to D along any available route. In this example, there are two possible routes toward D—one via a direct connection to C and the other via router E. The way the loose hop is picked depends on the IGP route that is available toward D.

Figure 6-9 *Explicit Route, Loose Hops*

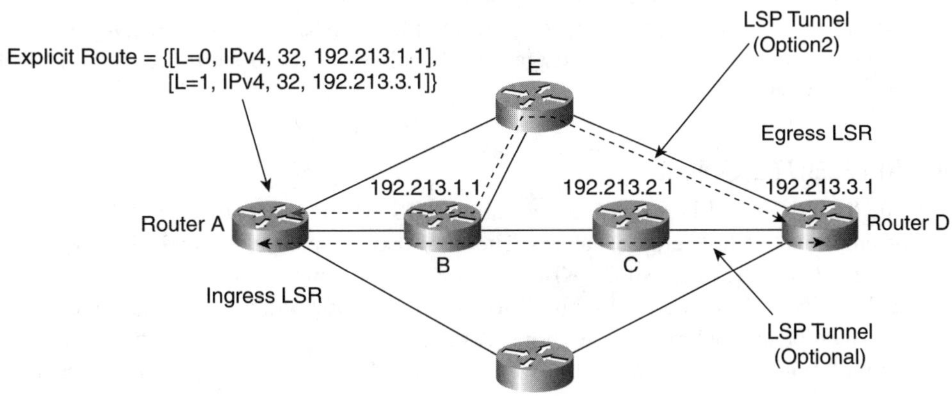

It is important to note that intermediate LSRs between the sender and receiver may also change the ERO by inserting subobjects. An example is where an intermediate node replaces a loose route subobject with a strict route subobject to force the traffic around a specific path. Also, the presence of loose nodes in an explicit route implies that it is possible to create forwarding loops in the underlying routing protocol during transients. Loops in an LSP tunnel can be detected using the RECORD_ROUTE object (RRO), as discussed in the next section.

RECORD_ROUTE Object

The RRO is used to collect detailed path information and is useful for loop detection and for diagnostics. By adding an RRO to the PATH message, the sender can receive information about the actual PATH taken by the LSP. Remember that although the ERO specifies an explicit PATH, the PATH might contain loose hops, and some intermediate nodes might change the ERO, so the final PATH recorded by the RRO could be different from the ERO specified by the sender. The RRO can be present in both RSVP PATH and RESV messages. The RRO is present in an RESV message if the RRO that has been recorded on the PATH message needs to be returned to the ingress LSR.

There are three possible uses of RROs in RSVP:

- **Loop detection**—An RRO can function as a loop-detection mechanism to discover L3 routing loops or loops inherent in the explicit route.
- **Path information collection**—An RRO collects up-to-date detailed path information hop-by-hop about RSVP sessions, providing valuable information to the sender or receiver. Any path change (because of network topology changes) is reported.
- **Feedback into ERO**—An RRO can be used as input to the ERO object. If the sender receives an RRO via the RESV message, it can alter its ERO in the next PATH message. This can be used to “pin down” a session path to prevent the path from being altered even if a better path becomes available.

The initial RRO contains only one subobject: the sender’s IP addresses. When a PATH message containing an RRO is received by an intermediate router, the router stores a copy of it in the path state block. When the PATH message is forwarded to the next hop, the router adds to the RRO a new subobject that contains its own IP address. When the receiver sends the RRO to the sender via the RESV message, the RRO has the complete route of the LSP from ingress to egress.

SESSION_ATTRIBUTE Object

The SESSION_ATTRIBUTE object allows RSVP-TE to set different LSP priorities, preemption, and fast-reroute features. These are used to select alternate LSPs in case of a failure in the network. The SESSION_ATTRIBUTE is carried in the PATH message. It includes fields such as Setup Priority and Holding Priority, which affect whether this session can preempt or can be preempted by other sessions. A Flag field is also used to introduce options such as whether transit routers can use local mechanisms that would violate the ERO and cause local repair. Other Flag options indicate that the tunnel ingress node may choose to reroute this tunnel without tearing it down.

FLOW_SPEC Object

An elementary RSVP reservation request consists of a FLOW_SPEC together with a FILTER_SPEC; this pair is called a *flow descriptor.* The FLOW_SPEC object specifies a desired QoS. The FILTER_SPEC object, together with a SESSION object specification, defines the set of data packets—the "flow"—to receive the QoS defined by the flowspec. An example of the use of FLOW_SPEC with RSVP-TE would be to indicate which path certain traffic gets put on based on the QoS characteristics of such traffic. Data packets that are addressed to a particular session but that do not match any of the filter specs for that session are handled as best-effort traffic. The flowspec in a reservation request generally includes a service class and two sets of numeric parameters:

- An Rspec (R for "reserve") that defines the desired QoS
- A Tspec (T for "traffic") that describes the data flow

SENDER_TEMPLATE Object

PATH messages are required to carry a SENDER_TEMPLATE object, which describes the format of data packets that this specific sender originates. This template is in the form of a FILTER_SPEC that is typically used to select this sender's packets from others in the same session on the same link. The extensions of RSVP for TE define a new SENDER_TEMPLATE C-Type (LSP_TUNNEL_IPv4) that contains the IPv4 address for the sender node and a unique 16-bit identifier, the LSP_ID, that can be changed to allow a sender to share resources with itself. This LSP_ID is used when an LSP tunnel that was established with an SE reservation style is rerouted.

SESSION Object

The SESSION object is added to the PATH message to help identify and diagnose the session. The new LSP_TUNNEL_IPv4 C-Type contains the IPv4 address of the tunnel's egress node and a unique 16-bit identifier that remains constant over the life of the LSP tunnel, even if the tunnel is rerouted.

Details of the RESV Message

An RESV message is transmitted from the egress LSR toward the ingress in response to the receipt of a PATH message. The RESV message is used for multiple functions, including: distributing label bindings, requesting resource reservations along the path, and specifying the reservation style (FF or SE).

The RSVP RESV message can contain a number of different objects such as LABEL, RECORD_ROUTE, SESSION, and STYLE.

Figure 6-10 shows the format of the RESV message.

Figure 6-10 *RSVP-TE RESV Message*

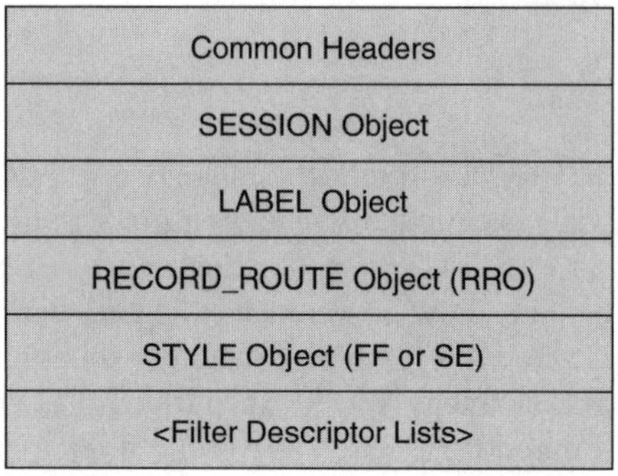

The RECORD_ROUTE and SESSION objects were described as part of the PATH message in the preceding section. The LABEL object contains the label or stack of labels that is sent from the downstream node to the upstream node. The STYLE object specifies the reservation style used. As you have learned, the FF and SE reservation styles filters are used for TE. For the FF and SE styles, a label is provided for each sender to the LSP.

Understanding MPLS Fast Reroute

One of the requirements for TE is the capability to reroute an established TE tunnel under various conditions. Such rerouting capabilities could include the following:

- Setting administrative policies to allow the LSP to reroute, such as when the LSP does not meet QoS requirements.
- Rerouting an LSP upon failure of a resource along the TE tunnel's established path.
- Setting an administrative policy that might require that an LSP that has been rerouted must return to its original path when a failed link or router becomes available.

Network operation must not be disrupted while TE rerouting is in progress. This means that you need to establish backup tunnels ahead of time and transfer traffic from the old tunnel to the new tunnel before you tear down the old tunnel. This concept is called *make-before-break.* A problem could arise if the old and new tunnels are competing for network resources; this might prevent the new tunnel from being established, because the old tunnel that needs to be torn down still has the allocated resources.

One of the advantages of using RSVP-TE is that the protocol has many hooks to take care of such problems. RSVP uses the SE reservation style to prevent the resources used by an old tunnel from being released until the new tunnel is established. The SE reservation style also prevents double counting of the resources when moving from an old tunnel to a new tunnel.

The speed of rerouting a failed tunnel is crucial for maintaining SLAs for real-time applications in the metro. When an LSP tunnel fails, the propagation of the failure to the ingress LSR/LER that established the tunnel and the convergence of the network to a new LSP tunnel could cause higher-level applications to time out. MPLS fast reroute allows an LSP tunnel to be rerouted in tens of milliseconds.

RSVP-TE can be used to establish backup LSP tunnels if active LSP tunnels fail. There are two methods of doing so:

- End-to-end repair
- Local repair

End-to-End Repair

With the end-to-end repair method, the whole LSP tunnel is backed up from the ingress LSR to the egress LSR. If the LSP fails because of a break in the network, a whole new LSP is established end to end. In this case, it is possible to presignal the secondary path, which is quicker than resignaling the LSP.

Local Repair

Local repair allows the LSP to be repaired at the place of failure. This allows the existing LSP to reroute around a local point of failure rather than establish a new end-to-end LSP. The benefit of repairing an LSP at the point of failure is that it decreases the network convergence time and allows the traffic to be restored in tens of milliseconds. This is important to meet the needs of real-time applications such as Voice over IP or video over IP, which are the next-generation services for metro networks.

To achieve restoration in tens of milliseconds, backup LSPs are signaled and established in advance of failure. The traffic is also redirected as close to the failure as possible. This reduces the delays caused by propagating failure notification between LSRs.

Figure 6-11 shows the difference between using local repair and end-to-end repair.

Figure 6-11 *The Value of Local Repair*

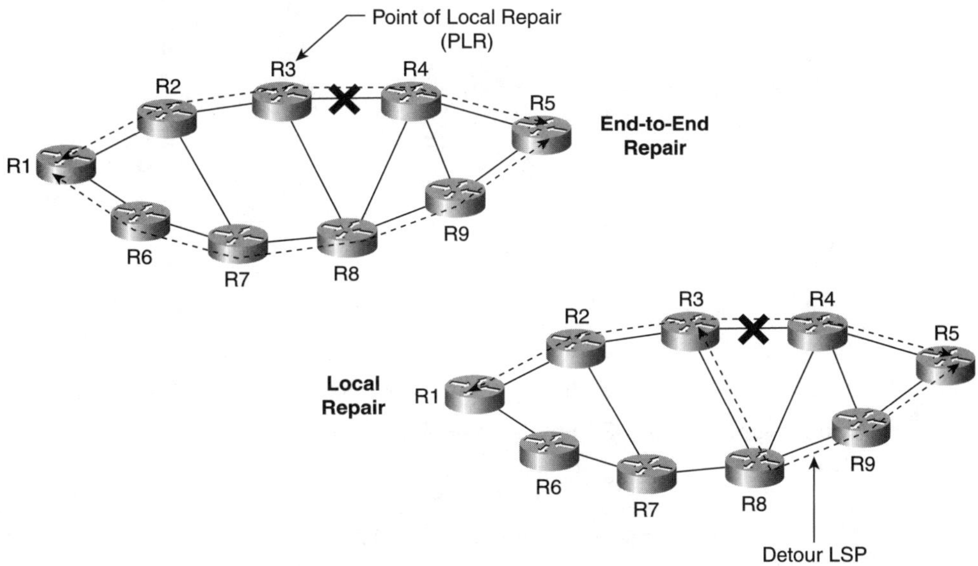

In Figure 6-11, an LSP tunnel is established between R1 and R5. If end-to-end repair is used and a failure occurs anywhere on the links or routers between R1 and R5—the R3-R4 link in this example—failure notification has to propagate from R3 all the way to R1. Also, all the LSRs, including R1 and R2, have to be involved in recomputing the new path. If the secondary path is presignaled between R1 and R5, convergence occurs much faster.

Conversely, local repair allows the traffic to be redirected closest to the failure and hence dramatically reduces the restoration time. If local repair is used, the LSP could be spliced between R3 and R5, bypassing the failure. Of course, this is all great as long as you know where the failure will occur so that you can work around it. Because this is impossible to know, you have to predict which links are carrying critical data and need to be protected. Two local repair techniques, one-to-one backup and facility backup, are discussed next.

One-to-One Backup

In the one-to-one backup method, a node is protected against a failure on its downstream link or node by creating an LSP that starts upstream of that node and intersects with the original LSP somewhere downstream of the point of link or node failure. In Figure 6-11 (local repair) the one-to-one backup method was used to protect against a failure of the link R3-R4, or the failure of node R4. In this case, R3's backup is an LSP that starts at R3 and ends downstream of the R3-R4 link on the R5 node. The partial LSP that starts from R3 and goes around R4 and splices back into the original LSP is called a *detour LSP.* To fully protect an LSP that passes N nodes, there could be as many as N–1 detours. In the example in Figure 6-12, to protect the LSP between R1 and R5, there could be as many as four detour LSPs.

Figure 6-12 *Full LSP Protection*

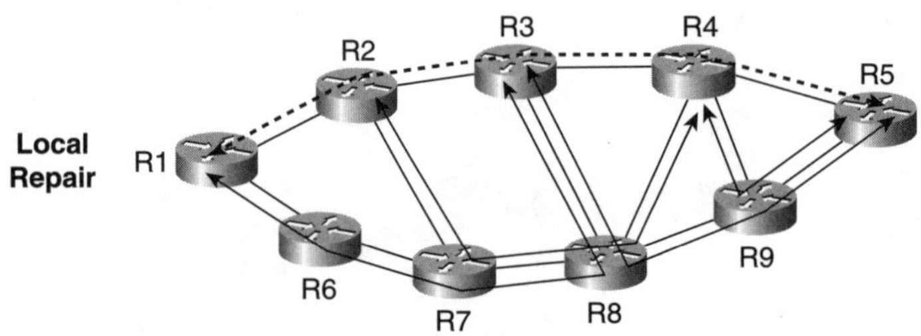

The LSP that needs to be protected is R1-R2-R3-R4-R5:

- Upon failure of the R1-R2 link, or R2 node, R1's detour LSP would be R1-R6-R7-R8-R3.
- Upon failure of the R2-R3 link, or R3 node, R2's detour LSP would be R2-R7-R8-R4.
- Upon failure of the R3-R4 link, or R4 node, R3's detour LSP would be R3-R8-R9-R5.
- Upon failure of the R4-R5 link, R4's detour LSP would be R4-R9-R5.

The point (router) that initiates the detour LSP is called the *point of local repair (PLR).* When a failure occurs along the protected LSP, the PLR redirects the traffic onto the local detour. If R1-R2 fails, R1 switches the traffic into the detour LSP R1-R6-R7-R8-R3.

Facility Backup—Bypass

Another method for protecting the LSP against failure is called the *facility backup.* Instead of creating a separate LSP for every backed-up LSP, a single LSP is created that serves to back up a set of LSPs. This LSP is called a *bypass tunnel.* The bypass tunnel intersects the path of the original LSPs downstream of the PLR. This is shown in Figure 6-13.

Figure 6-13 *Bypass Tunnel*

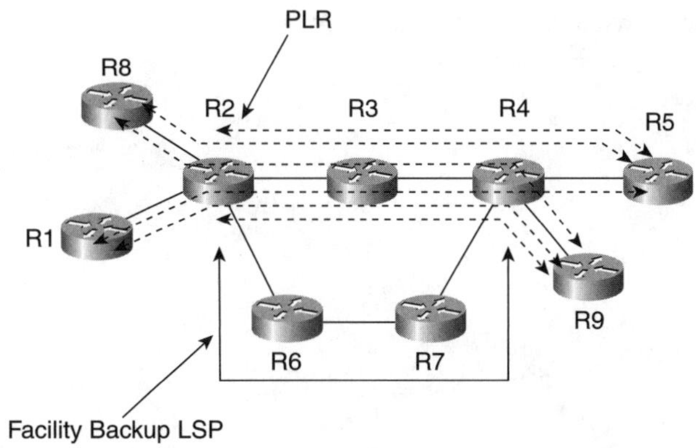

The bypass tunnel R2-R6-R7-R4 is established between R2 and R4. The scalability improvement from this technique comes from the fact that this bypass tunnel can protect any LSP from R1, R2, and R8 to R4, R5, and R9. As with the one-to-one technique, to fully protect an LSP that traverses N nodes, there could be as many as N–1 bypass tunnels. However, each of these bypass tunnels can protect a set of LSPs.

Conclusion

This chapter has discussed the basics of RSVP-TE and how it can be applied to establish LSPs, bandwidth allocation, and fast-reroute techniques. A detailed explanation of the RSVP-TE messages and objects was offered to give you a better feel for this complex protocol, which probably requires a book of its own. Many of the techniques explained in this chapter apply to provisioning scalable L2 metro Ethernet services.

The metro will consist of a mix of technologies ranging from Ethernet switches to SONET/SDH equipment to optical switches. Creating a unified control plane that is capable of provisioning LSP tunnels end to end and helping in the configuration and management of such equipment becomes crucial. You have seen the MPLS control plane used for packet networks. The flexibility and standardization of MPLS is extending its use to TDM and optical networks. The next two chapters discuss Generalized MPLS (GMPLS) and how this control plane becomes universal in adapting not only to packet networks but also across TDM and optical networks.

This chapter covers the following topics:

- Understanding GMPLS
- Establishing the Need for GMPLS
- Signaling Models
- Label Switching in a Nonpacket World

CHAPTER 7

MPLS Controlling Optical Switches

The operation of today's optical networks is manual and operator-driven, which increases network operational complexities and cost. The industry has been looking for methods that reduce the operational burden of manual circuit provisioning, reduce costs, and offer a more dynamic response to customer requirements. In other words, the industry wants to be able to deploy time-division multiplexing (TDM) and optical circuits more dynamically and wants faster provisioning times.

The principles upon which MPLS technology is based are generic and applicable to multiple layers of the transport network. As such, MPLS-based control of other network layers, such as the TDM and optical layers, is also possible. The Common Control and Measurement Plane (CCAMP) Working Group of the IETF is currently working on extending MPLS protocols to support multiple network layers and new TDM and optical services. This concept, which was originally referred to as Multiprotocol Lambda Switching (MPλS), is now referred to as Generalized MPLS (GMPLS). This chapter refers to definitions from the CCAMP Working Group in the areas that cover the GMPLS architecture and concepts.

Understanding GMPLS

Generalized MPLS is a set of architectures and protocols that enables TDM and optical networks to behave more dynamically. GMPLS builds on the MPLS control, which is well known and proven to work, and extends the capabilities of MPLS to control TDM and optical networks, including TDM switches, wavelength switches, and physical port switches.

In the same way that MPLS builds label switched paths (LSPs) between packet switches, GMPLS extends the concept of LSPs to TDM and optical switches. Figure 7-1 illustrates a three-layer hierarchy where GMPLS LSPs are built between two points in the network over multiple layers.

Figure 7-1 shows how the MPLS LSP concept that is used for the IP packet/cell layer can be extended to address the TDM and optical layers. On the IP layer, an LSP is formed between routers A and I. On the TDM layer, an LSP is formed between SONET/SDH multiplexers J and N. On the photonics layer, an LSP is formed between optical switches

S and W along the path S-T-U-V-W. The establishment of LSPs of course necessitates that TDM and optical switches become aware of the GMPLS control plane while still using their own multiplexing and switching techniques. This is one of the powerful advantages of MPLS, because the control and forwarding planes are decoupled.

Figure 7-1 *GMPLS LSPs*

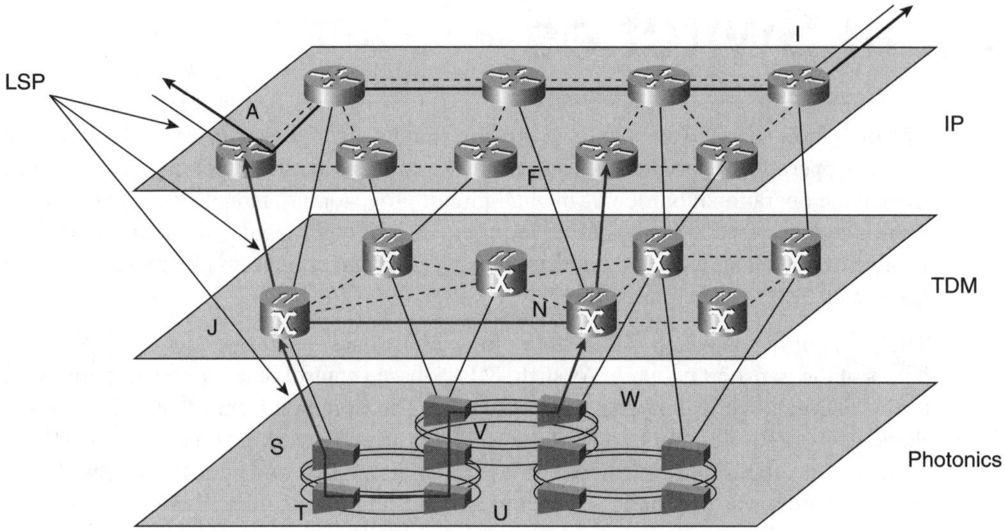

GMPLS has two applications, both of which can be used in metro network deployments. First, for dynamic circuit provisioning, GMPLS can be used to establish point-to-point or multipoint-to-point virtual private optical networks. Second, GMPLS can be used for protection on the circuit level. In the context of deploying Ethernet services over an optical cloud, GMPLS would extend across L2 Ethernet switches/routers, SONET/SDH multiplexers, and optical cross-connects (OXCs) to establish end-to-end circuits. Note that such deployments have not occurred yet, and it is unclear at the moment how fast or slow the adoption of GMPLS will evolve. The next section describes in more detail the need for GMPLS in optical networks.

Establishing the Need for GMPLS

Anyone who has been in the networking industry for a while would likely raise the issue of whether GMPLS is really needed or is overkill. After all, we have managed so far to build large-scale TDM networks with all sorts of methods, and we have seen improvement in tools to facilitate the operation and management of those networks. To understand the issue of whether GMPLS is necessary, you need to look first at how TDM networks function today and then at how they could benefit from GMPLS.

The following section describes the provisioning model of today's network deployments, which are more static with centralized management. The problem with this model is that it doesn't enable carriers to provide new services that involve the dynamic establishment and restoration of TDM and optical circuits while minimizing the operational cost and provisioning times. This is the problem that the GMPLS model attempts to address. If GMPLS could solve this problem, the result would be a better service experience for customers and increased revenue for the carrier. However, adopting GMPLS would also require fundamental changes to the way you administer, manage, and build networks.

Static and Centralized Provisioning in TDM Networks

Currently, TDM and optical networks are statically provisioned. Provisioning a point-to-point circuit takes weeks to accomplish, because it entails lengthy administrative and architectural tasks. The majority of today's TDM network management and provisioning models are centralized. Provisioning is done either manually or with automated tools and procedures that reside in a central network management entity that has knowledge of the whole network and its elements. To handle scalability issues, such as having too many nodes (thousands) to manage, network managers use a hierarchical approach in which they manage multiple domains separately and higher management layers oversee the whole service operation. The network topology includes topology information about rings and meshed networks. The network resources include information about the network elements, such as fibers, ducts, links, and their available capacity. Entering such information manually is tedious and error-prone, especially in networks that require constant changes for expansion and upgrades.

The provisioning process involves the following:

- **Administrative tasks**—Request for a circuit involves the paperwork or web-based tools for a customer such as a large enterprise to fill out and submit as a request for a circuit. The request is fulfilled by the network operator.
- **Network planning**—The network operator has to run simulations to find out whether the network has the capacity to absorb the additional circuits and to determine how to optimize the network resources. This task is normally done on a set of circuits at regular intervals. Network planning touches different parts of the network, depending on where these circuits start and end. High-capacity circuits normally put a major strain on metropolitan area networks that were built for traditional voice services. In some cases, the addition of one TDM circuit might cause the operator to build more metro SONET rings to absorb additional capacity; hence, the service could be rejected or delayed until the operator justifies the economics of building more circuits for a particular customer.
- **Installing the physical ports**—This is the manual task of installing the WAN ports at the customer premises and installing the connection/circuit between the customer premises and the operator's networks.

- **Circuit provisioning**—This is the task of establishing the circuit end to end, using either management tools or manual configuration. Circuit provisioning is one of the most challenging areas because it requires establishing circuits across multiple components, sometimes from different vendors, with different interfaces and different protocols. Circuit provisioning also involves testing the circuit to see whether it complies with the SLA that was promised to the customer.
- **Billing**—As simple as it may sound, a service cannot be deployed until it can be billed for. Whether flat billing or usage-based billing is used, the task of defining and accounting for the right variables is not simple.
- **Network management**—Last but not least is the continuous process of managing the different network elements, keeping the circuits up and running, and restoring the circuits in case of network failures.

The Effect of a Dynamic Provisioning Model

GMPLS offers a dynamic provisioning model for building optical networks. In this more dynamic and decentralized model, information about the network topology and resources can be exchanged via protocols such as OSPF traffic engineering (OSPF-TE) and IS-IS traffic engineering (ISIS-TE). The information is available to all nodes in the network, including the Network Management System (NMS), which can act upon it. The dissemination of such information via the routing protocols gives the operator a clearer view of the network, which facilitates planning, provisioning, and operation.

Figures 7-2 and 7-3 show two scenarios of centralization and decentralization.

Figure 7-2 *Centralized, Static Control*

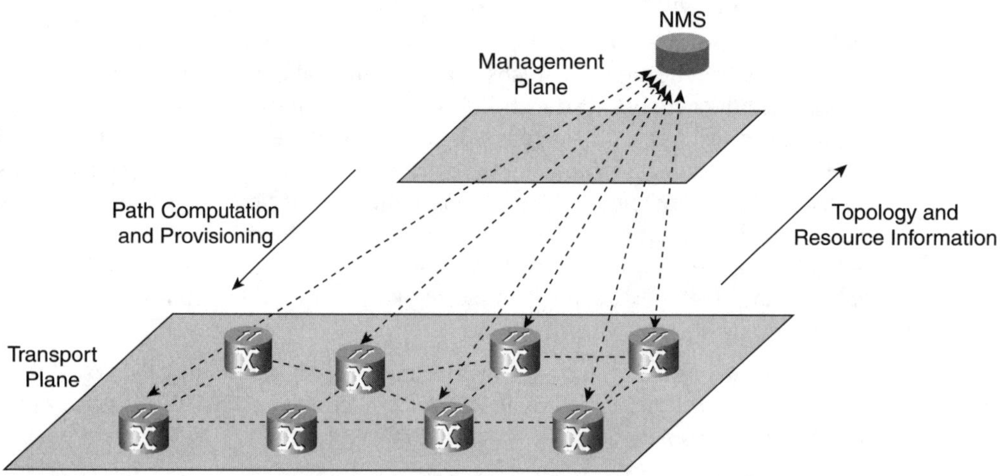

Figure 7-2 shows the centralized approach, in which all nodes communicate with the NMS and relay information about topology and resources to a central database. The NMS acts on this information for path computation and provisioning.

In optical networks, the control plane can be exchanged between the different network systems (optical switches, routers, and so on) via in-band or out-of-band communications. The GMPLS control plane can use multiple communication models:

- Over a separate fiber
- Over a separate wavelength
- Over an Ethernet link
- Over an IP tunnel through a separate management network
- Over the overhead bytes of the data-bearing link

A communication over a SONET/SDH data communication channel (DCC), for example, could use the SONET/SDH DCC path D1-D3 or the line D4-D24 overhead bytes. For wavelength-division multiplexing (WDM) nodes, a separate wavelength could be dedicated as an IP management channel. It is important that the management channel be operational at all times. If, for example, the management is done in-band, a network failure could cause the management channel to fail. Hence, the nodes and links could become inaccessible and couldn't be restored.

Figure 7-3 shows the decentralized approach, in which the nodes exchange topology and resource information via different protocols (for example, OSPF-TE) through the IP control plane running in-band or out-of-band. Path computation and provisioning can be triggered dynamically or via an NMS station. The NMS station could simply send commands to one of the ingress nodes to initiate a path.

Figure 7-3 *Decentralized, Dynamic Control*

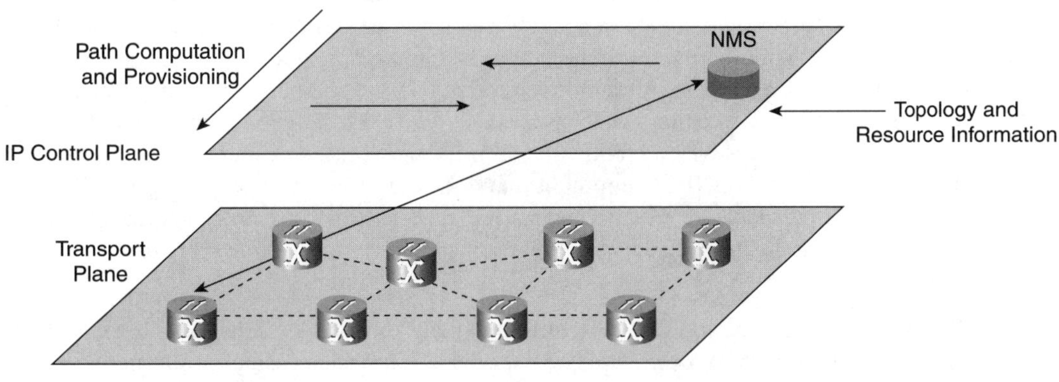

When applying routing to circuit-switched networks, it is useful to compare and contrast this situation with the IP packet routing case, which includes the following two scenarios:

- Topology and resource discovery
- Path computation and provisioning

Topology and Resource Discovery

In the case of routing IP packets, all routes on all nodes must be calculated exactly the same way to avoid loops and "black holes." Conversely, in circuit switching, routes are established per circuit and are fixed for that circuit. To accommodate the optical layer, routing protocols need to be supplemented with new information, such as available link capacity. Due to the increase in information transferred in the routing protocol, it is important to separate a link's relatively static parameters from those that may be subject to frequent changes.

Using a dynamic model to report link capacity in TDM and optical networks can be challenging. You have to find a balance where you are getting accurate reports about specific signals without flooding the network with too much information.

Path Computation and Provisioning

In packet networks, path computation and reachability are very dynamic processes. Routing protocols determine the best path to a destination based on simple metrics such as link bandwidth. As described in Chapter 6, "RSVP for Traffic Engineering and Fast Reroute," MPLS with RSVP-TE gives you more control to traffic-engineer the network. For optical networks, path computation and provisioning depend on the following information:

- The available capacity of the network links
- The switching and termination capabilities of the nodes and interfaces
- The link's protection properties

When such information is exchanged dynamically via routing protocols, the network always has a real-time view of link and node capacity and properties that can be used to calculate the most suitable path.

With all the required tasks for deploying a service, optimizing the right mix of tasks becomes challenging. No one solution has a positive impact on all variables at the same time. Applying a dynamic provisioning model to the network, for example, would shorten provisioning but would also make network planning, service billing, and network management more challenging. After all, carriers have always dealt with a static provisioning and TE model, because they have always had total control of the network, its resources, and its behavior. Besides, for legacy SONET equipment that does not have the capability to run GMPLS and dynamic protocols, static approaches remain necessary. As such, a combination of static and dynamic, centralized and decentralized approaches would apply to most network designs.

The transition to adopting GMPLS will take many steps and will happen faster with some providers than others. Adopting this new model will be much easier for alternative providers and greenfield operators than for incumbents, which have well-defined procedures and tools that have been used for years. The cost justification for adopting GMPLS is not yet as clear as its benefits are. The next section discusses the dynamic provisioning model in more detail.

To adapt MPLS to control TDM and optical networks, the following primary issues need to be addressed:

- Addressing
- Signaling
- Routing
- Restoration and survivability

The following section begins by looking at the different signaling models that are in use and that are proposed for optical networks. Chapter 8, "GMPLS Architecture," provides more details about the rest of the topics in the preceding list.

Signaling Models

Signaling is a critical element in the control plane. It is responsible for establishing paths along packet-switched capable (PSC) and non-PSC networking devices such as routers, TDM cross-connects, and OXCs. PSC networks have no separation between the data and signaling paths; both data traffic and control traffic are carried over the same channels. In optical networks, control traffic needs to be separated from data traffic. One of the reasons is that OXCs are transparent to the data, because they perform light or lambda switching, whereas control traffic needs to be terminated at each intermediary OXC, because it carries the information to manage the data flows and information exchange between OXCs.

Multiple proposals exist for a signaling infrastructure over optical networks. The most common models are the following:

- The overlay model
- The peer model
- The augmented model

The Overlay Model

In this model, illustrated in Figure 7-4, the internals of the optical infrastructure are totally transparent to the data-switching infrastructure. The optical infrastructure is treated as a separate intelligent network layer. Data switches at the edges of the optical infrastructure can statically or dynamically provision a path across the optical cloud. This is very similar to the IP-over-ATM model that exists today in carrier backbones. In this model, two independent control planes exist:

- **Within the packet layer**—The control plane runs on the User-to-Network Interface (UNI) between the data switches at the edge of the optical cloud and the optical switches.
- **Within the optical network**—The control plane runs on the Network-to-Network Interface (NNI) between the optical switches.

Figure 7-4 *Overlay Model*

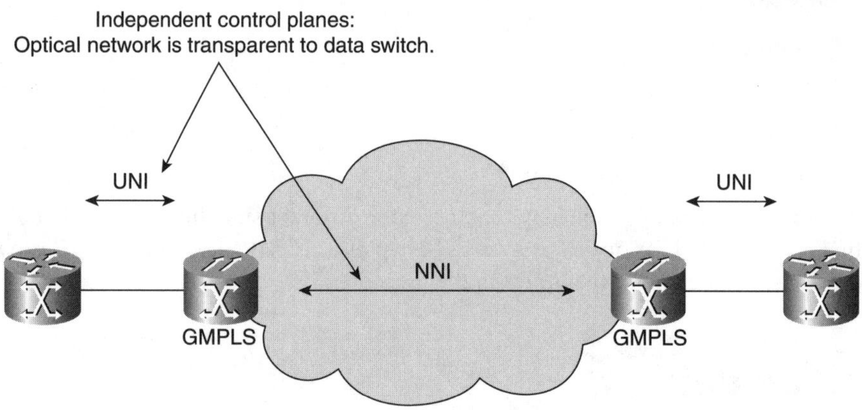

The overlay model applies in environments with limited or unknown trust that apply strict levels of policy and authentication and that limit routing information transfer.

The Peer Model

In the peer model, illustrated in Figure 7-5, the IP/MPLS layers act as peers of the optical transport network, such that a single control plane runs over both the IP/MPLS and optical domains. As far as routing protocols are concerned, each edge device is adjacent to the optical switch it is attached to. The label switch routers (LSRs) and OXCs exchange complete information. The routers/data switches know the full optical network topology and can compute paths over it. For data-forwarding purposes, a full optical mesh between edge devices is still needed so that any edge node can communicate with any other edge node.

Figure 7-5 *Peer Model*

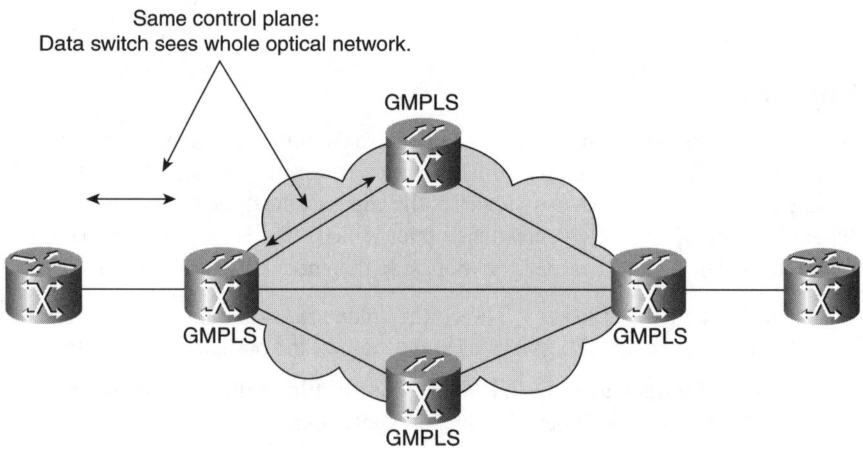

The advantage of the peer model is that, by developing uniform control, it gives the IP layer visibility into the optical layer and supports better IGP scaling if routers are meshed over an operational network. The peer model is much more similar to the use of MPLS than is an IP-over-ATM overlay model.

The Augmented Model

The augmented model, illustrated in Figure 7-6, is a hybrid model that falls between the overlay and peer models. In the augmented model, separate control planes for the optical and IP domains are used, but some edge data switches still could have a limited exchange of routing information with border optical switches. This model allows for a transition from the overlay model to the more evolved peer model. One possible scenario in which the augmented model could be used is where a provider owns the data switches and the border optical switches and relies on a transport service offered by a different provider that owns the core optical switches.

Figure 7-6 *Augmented Model*

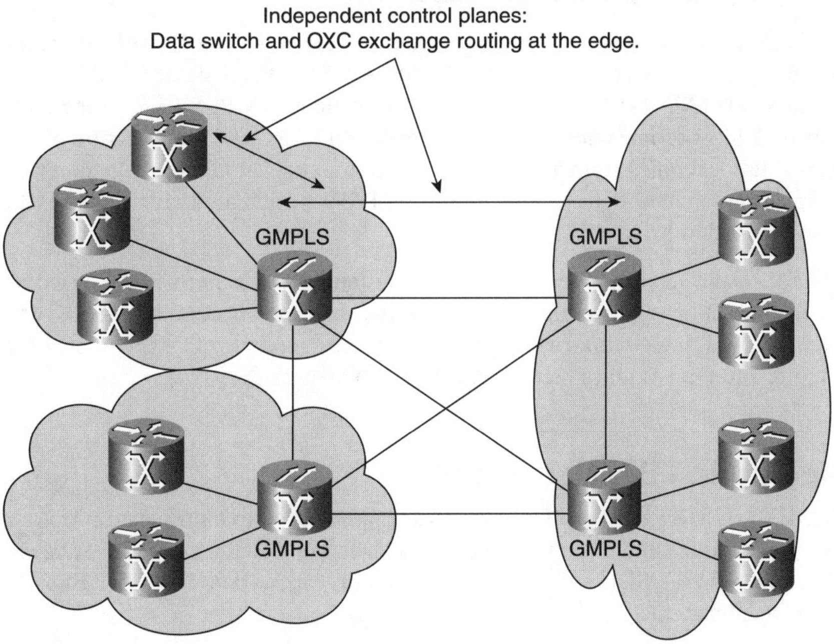

Label Switching in a Nonpacket World

MPLS networks consist of LSRs connected via circuits called label switched paths (LSPs). To establish an LSP, a signaling protocol is required. Between two adjacent LSRs, an LSP is locally identified by a short, fixed-length identifier called a *label*, which is only significant between these two LSRs. When a packet enters an MPLS-based packet network, it is classified according to its forwarding equivalency class and, possibly, additional rules, which together determine the LSP along which the packet must be sent. For this purpose, the ingress LSR attaches an appropriate label to the packet and forwards the packet to the next hop. The label itself is a shim layer header, a virtual path identifier/virtual channel identifier (VPI/VCI) for ATM, or a data-link connection identifier (DLCI) for Frame Relay. When a packet reaches a core packet LSR, that LSR uses the label as an index into a forwarding table to determine the next hop, and the corresponding outgoing label. The LSR then writes the new label into the packet and forwards the packet to the next hop. When the packet reaches the egress LSR (or the one node before the egress LSR for penultimate hop popping), the label is removed and the packet is forwarded using appropriate forwarding, such as normal IP forwarding.

So how do these concepts apply to networks that are not packet-oriented, such as TDM- and WDM-based networks?

In TDM networks, the concept of label switching happens at the circuit level or segment level. Switching can happen, for example, at the time-slot level where an input OC3 time slot is cross-connected to an output OC3 time slot.

For WDM-capable nodes, switching happens at the wavelength level, where an input wavelength is cross-connected to an output wavelength. As such, SONET/SDH add/drop multiplexers (ADMs) and OXCs become equivalent to MPLS LSRs, time-slot LSPs and lambda LSPs become equivalent to packet-based LSPs, and the selection of time slots and wavelength becomes equivalent to the selection of packet labels. Also, nonpacket LSPs are bidirectional in nature, in contrast to packet LSPs, which are unidirectional (this is covered in more depth in Chapter 8).

The following section takes a closer look at label switching in TDM-based networks and touches upon label switching in WDM networks. The concepts of label switching in both TDM and WDM networks are similar in the sense that with TDM networks GMPLS controls circuits and with WDM GMPLS controls wavelengths.

Label Switching in TDM Networks

SONET and SDH are two TDM standards that are used to multiplex multiple tributary signals over optical links, thus creating a multiplex structure called the *SONET/SDH multiplex.* Details about the SONET/SDH structure are covered in Appendix A, "SONET/SDH Basic Framing and Concatenation."

If you choose to use the GMPLS control plane to control the SONET/SDH multiplex, you must decide which of the different components of the SONET/SDH multiplex that can be switched need to be controlled using GMPLS. As described in Appendix A, the SONET/SDH frame format consists of overhead bytes, a payload, and a pointer to the payload. Essentially, every SONET/SDH element that is referenced by a pointer can be switched. These component signals in the SONET case are the synchronous transport signal (STS), Synchronous Payload Envelopes (SPEs), and virtual tributaries (VTs), such as STS-1, VT-6, VT-3, VT-2, and VT-1.5. For SDH, the elements that can be switched are the VC-4, VC-3, VC-2, VC-12, and VC-11.

When concatenation is used in the case of SONET or SDH, the new structure can also be referenced and switched using GMPLS. As explained in Chapter 2, "Metro Technologies," concatenation—standard or virtual—allows multiple tributaries or STS/STM to be bonded to create a bigger pipe. GMPLS can be applied on the concatenated pipe.

The following sections discuss in more detail the concepts of label switching in a TDM network, including the following:

- Signaling in a TDM network
- SONET/SDH LSRs and LSPs
- The mechanics and function of a TDM label

Signaling in a TDM Network

To support signaling in the TDM network, several modifications need to be made to MPLS. First, the traditional MPLS label needs to be modified to provide better binding between the label itself and the circuit it represents on a particular interface. Second, an LSP hierarchy needs to be introduced so that LSPs that represent signals can be tunneled inside other LSPs. Third, the capabilities of the label distribution protocols need to be extended so that they can distribute the information that is necessary to switch the signals along the path. A high-level description of the signaling modifications is covered in the next section, and a more detailed description is available in Chapter 8.

SONET/SDH LSRs and LSPs

GMPLS defines a SONET/SDH terminal multiplexer, an ADM, and a SONET cross-connect as SONET/SDH LSRs. A path or circuit between two SONET/SDH LSRs becomes an LSP. A SONET/SDH LSP is a logical connection between the point at which a tributary signal (client layer) is adapted to its SPE for SONET or to its virtual container for SDH, and the point at which it is extracted from its SPE or virtual container. Figure 7-7 shows a SONET/SDH LSP. In this example, an STS-1 LSP is formed between path terminal equipment—PTE1 and PTE2—across line terminal equipment—LTE1 and LTE2. The LTEs are the SONET/SDH network elements that originate or terminate the line signal. The PTEs are the SONET/SDH network elements that multiplex/demultiplex the payload. A PTE, for example, would take multiple DS1s to form an STS-1 payload.

Figure 7-7 *GMPLS LSP Across SONET Equipment*

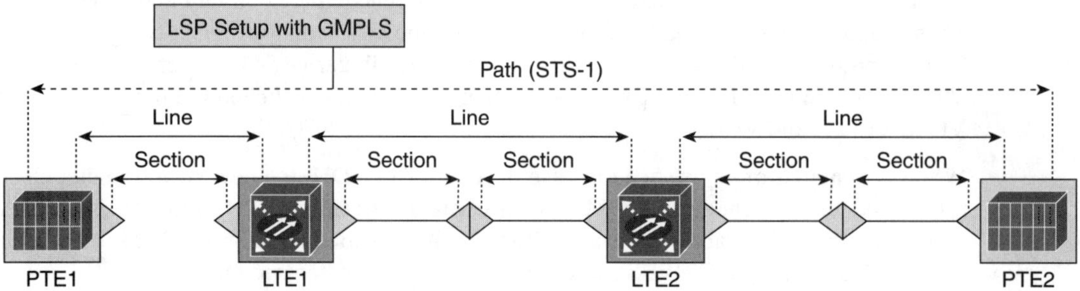

To establish a SONET/SDH LSP, a signaling protocol is required to configure the input interface, switch fabric, and output interface of each SONET/SDH LSR along the path. A SONET/SDH LSP can be point-to-point or point-to-multipoint, but not multipoint-to-point, because no merging is possible with SONET/SDH signals. To facilitate the signaling and setup of SONET/SDH circuits, a SONET/SDH LSR must identify each possible signal individually per interface, because each signal corresponds to a potential LSP that can be established through the SONET/SDH LSR. GMPLS switching does not apply to all possible SONET/SDH signals—only to those signals that can be referenced by a SONET/SDH pointer, such as the STS SPEs and VTs for SONET and the VC-*X*s for SDH.

The next section addresses the mechanics and functions of a GMPLS label in the context of TDM networks.

The Mechanics and Function of a TDM Label

You have already seen label switching adopted with an asynchronous technology such as IP where a label attaches to an IP packet and helps put that packet on the right LSP in the direction of its destination. For SONET/SDH, which are synchronous technologies that define a multiplexing structure, GMPLS switching does not apply to individual SONET/SDH frames. GMPLS switching applies to signals, which are continuous sequences of time slots that appear in a SONET/SDH frame. GMPLS can switch SONET/SDH signals. As such, a SONET/SDH label needs to indicate the signals that can be switched, such as the STS SPE, VTs, and virtual containers.

Figure 7-8 compares label switching applied to TDM and traditional label switching in the packet world.

As you can see, with a packet LSR, the labels are identified for a certain forwarding equivalency class and are used to label-switch the packet to its destination. The labels themselves are carried inside the IP packets for the LSR to perform the label-switching function. In the case of a

SONET/SDH LSR, the GMPLS control plane needs to map labels for the signals that need to be switched on each interface. In this example, the STS-1 signal on interface I1 is mapped to label 10 and is cross-connected to the STS-1 signal on interface I3, which is mapped to label 30. The VT 1.5 signal on interface I2 is mapped to label 20 and is cross-connected to the VT 1.5 signal on interface I4, which is mapped to label 40. Note that the SONET/SDH frames themselves do not carry any label; the mapping is just an indication by the GMPLS plane to allow the SONET/SDH node to perform the required switching function of the appropriate signals.

Figure 7-8 *SONET/SDH Label Switching*

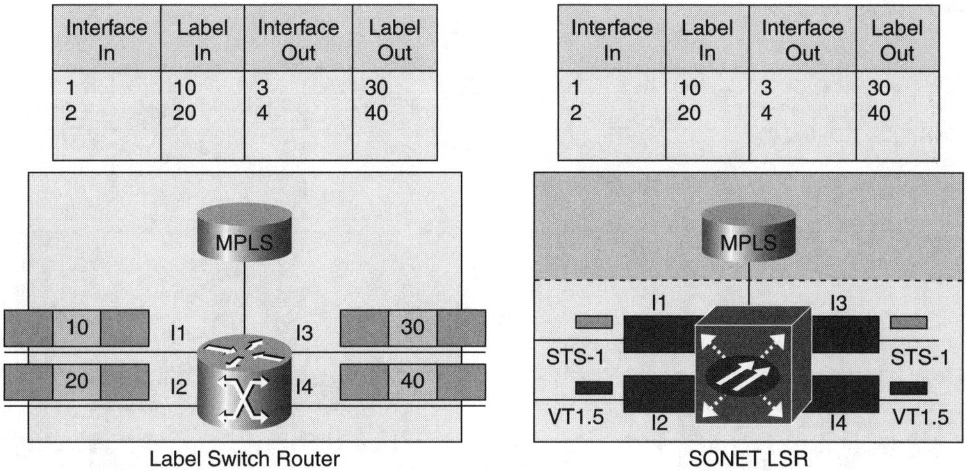

A SONET/SDH LSR has to identify each possible signal individually per interface to fulfill the GMPLS operations. To stay transparent, the LSR obviously should not touch the SONET/SDH overheads; this is why an explicit label is not encoded in the SONET/SDH overheads. Rather, a label is associated with each individual signal and is locally unique for each signal at each interface.

Because the GMPLS label is not coded in the signal itself, a mechanism needs to be established to allow the association of a label with SONET/SDH signals. The GMPLS label is defined in a way that enables it to give information about the SONET/SDH multiplex, such as information about the particular signal and its type and position in the multiplex.

Label Switching in WDM Networks

WDM is a technology that allows multiple optical signals operating at different wavelengths to be multiplexed onto a single fiber so that they can be transported in parallel through the fiber. OXCs in turn cross-connect the different wavelengths, in essence creating an optical path from source to destination. The optical path itself can carry different types of traffic, such as SONET/SDH, Ethernet, ATM, and so on. OXCs can be all optical, cross-connecting the wavelengths in the optical domain, or they can have optical-to-electrical-to-optical conversion, which allows for wavelength conversion mechanisms. In the GMPLS context, OXCs would run the GMPLS control plane and would become comparable to LSRs. Lambda LSPs are considered similar to packet-based LSPs, and the selection of wavelengths and OXC ports is considered similar to label selection.

Figure 7-9 compares the concept of MPLS switching in a WDM network in the same way that Figure 7-8 did for the TDM network.

Figure 7-9 *WDM Label Switching*

Interface In	Label In	Interface Out	Label Out
1	10	3	30
2	20	4	40

Interface In	Label In	Interface Out	Label Out
1	1	3	1
2	2	4	4

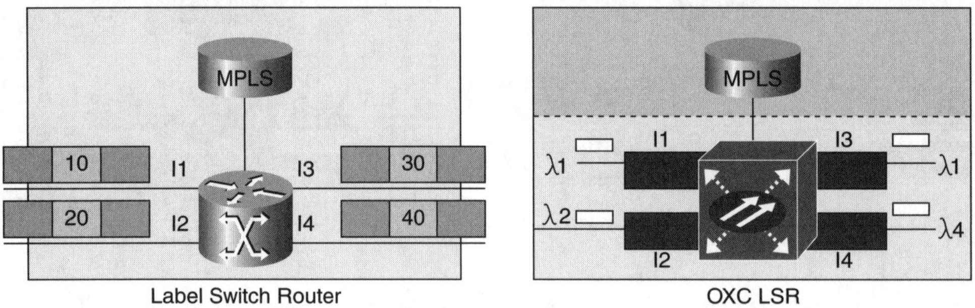

As already shown in the TDM example, the GMPLS labels are not carried inside the actual packet. In the case of an OXC LSR, the GMPLS control plane needs to map labels for the lambdas that need to be switched on each interface. In this example, label 1 on interface I1 is mapped to lambda 1 and cross-connected to lambda 1 on I3, which is mapped to label 1. Again, because the GMPLS label is not coded in the wavelength, a mechanism needs to be established to associate lambdas with labels. This is discussed in Chapter 8.

Conclusion

As discussed in this chapter, GMPLS is necessary to establish a dynamic way to provision optical networks. You have seen the benefits and drawbacks of both the static centralized and dynamic decentralized provisioning models. The chapter also discussed the different signaling models, such as the overlay, peer, and augmented models. These resemble how IP packet-based networks are deployed today over ATM or Frame Relay circuit-based networks. You have also seen how GMPLS uses labels to cross-connect the circuits for TDM and WDM networks. Although the concept of labels was adopted, the use of these labels is quite different from the traditional use of labels in data forwarding.

The next chapter goes into more detail about the extensions to routing and signaling that were added to the traditional MPLS control plane to accommodate optical networks.

This chapter covers the following topics:

- GMPLS Interfaces
- Modification of Routing and Signaling
- Inclusion of Technology-Specific Parameters
- Link Management Protocol
- GMPLS Protection and Restoration Mechanisms
- Summary of Differences Between MPLS and GMPLS

CHAPTER 8

GMPLS Architecture

Optical networks present some added challenges that do not normally exist in packet-switched networks (PSNs) and hence cannot be fully addressed by the traditional MPLS schemes. Here are a few examples of these challenges:

- Optical/TDM bandwidth allocation is done in discrete amounts, whereas in PSNs, bandwidth can be allocated from a continuous spectrum.
- The number of links in an optical network can be orders of magnitude larger than in a traditional network, due to the possible explosion in the number of parallel fibers deployed and the number of lambdas on each fiber. This in turn raises the issues of IP address assignment for optical links and the manageability of connecting ports on different network elements. If a fiber has 32 wavelengths, for example, between points A and B, and if each wavelength is treated as a separate link with its own addressing, the one fiber will create 32 different networks that need to be addressed and managed.
- Fast fault detection and isolation have always been advantages that optical networks have over PSNs.
- The fact that user data in an optical network is transparently switched necessitates the decoupling of user data from control plane information.

Generalized MPLS (GMPLS) attempts to address these challenges by building on MPLS and extending its control parameters to handle the scalability and manageability aspects of optical networks. This chapter explains the characteristics of the GMPLS architecture, such as the extensions to routing and signaling and the addition of technology parameters, that GMPLS adds to MPLS to be able to control optical networks.

GMPLS Interfaces

The GMPLS architecture extends MPLS to include five different types of interfaces used on label switch routers:

- **Packet-switch capable (PSC) interfaces**—Interfaces that can recognize packet boundaries and forward data based on packet headers. This is typical of interfaces on routers and L3 Ethernet switches.

- **Layer 2–switch capable (L2SC) interfaces**—Interfaces that can recognize L2 cell or frame boundaries and forward data based on L2 headers. This is typical of interfaces on ATM switches, Frame Relay switches, and L2 Ethernet switches.
- **Time-division multiplexing (TDM) interfaces**—Interfaces that can recognize time slots and forward data based on the data's time slot in a repeating cycle. This is typical of interfaces on digital cross-connects (DACSs), SONET add/drop multiplexers (ADMs), and SONET cross-connects. Such interfaces are referred to as TDM capable.
- **Lambda-switch capable (LSC) interfaces**—Interfaces that can forward data based on the lambda (wavelength) it was received on. This is typical of optical cross-connects (OXCs) that switch traffic on the wavelength level.
- **Fiber-switch capable (FSC) interfaces**—Interfaces that can forward data based on the position of the data in real-world physical spaces. This is typical of OXCs that switch traffic on the fiber or multiple-fiber level.

Modification of Routing and Signaling

The development of GMPLS requires the modification of current routing and signaling protocols. The adoption of a common, standardized control plane for managing packet/cell switches and optical switches is extremely important to the networking industry. This introduces a unified method for achieving fast provisioning, restoration, routing, monitoring, and managing data-switched and optical-switched networks while maintaining interoperability between multiple vendors. The MPLS control plane is being extended from controlling data switches to a more generic role of controlling any type of switching, including optical switching—hence the term Generalized MPLS.

To help MPLS span switches that are not packet-oriented, GMPLS introduces some modifications to MPLS in the areas of routing and signaling. The modifications take place in the following areas:

- Enhancements to routing protocols
- Enhancements to signaling protocols

The following sections discuss routing and signaling enhancements.

Enhancements to Routing

Introducing routing into TDM and optical networks does not mean turning TDM and optical nodes into IP routers, but rather using the benefits of routing protocols as far as relaying paths and resource information to better use network resources. In optical and TDM networks, this information includes the following:

- The available capacity of the network links
- The switching and termination capabilities of the nodes and interfaces
- The link's protection properties

This information is carried inside routing protocols such as Open Shortest Path First for Traffic Engineering (OSPF-TE) and IS-IS Traffic Engineering (IS-IS–TE). GMPLS introduces extensions to OSPF-TE and IS-IS–TE to allow these protocols to tailor to the specific information required by these networks. OSPF-TE and IS-IS–TE are extensions of the OSPF and IS-IS routing protocols that allow them to carry network information about available network resources. This information is used by protocols such as RSVP-TE to engineer the traffic in the network.

An MPLS TE link is considered to be like any regular link, meaning a link where a routing protocol adjacency is brought up via protocols such as OSPF. The link's Shortest Path First (SPF) properties and the TE properties are calculated and advertised. For GMPLS to accommodate optical networks, a few variations need to be introduced:

- Nonpacket links can be brought up without establishing a routing adjacency.
- A label switched path (LSP) can be advertised as a point-to-point TE link, and the advertised TE link need no longer be between two OSPF/IS-IS direct neighbors.
- A number of links can be advertised as a single TE link, and there is no one-to-one association between routing adjacencies and a TE link.

A GMPLS TE link has special TE properties that can be configured or obtained via a routing protocol. An example of TE properties would be the bandwidth accounting for the TE link, including the unreserved bandwidth, the maximum reservable bandwidth, and the maximum LSP bandwidth. Other properties include protection and restoration characteristics.

IS-IS–TE and OSPF-TE explain how to associate TE properties to regular (packet-switched) links. GMPLS extends the set of TE properties and also explains how to associate TE properties with links that are not packet-switched, such as links between OXCs.

Figure 8-1 shows a TE link.

Figure 8-1 *GMPLS TE Link*

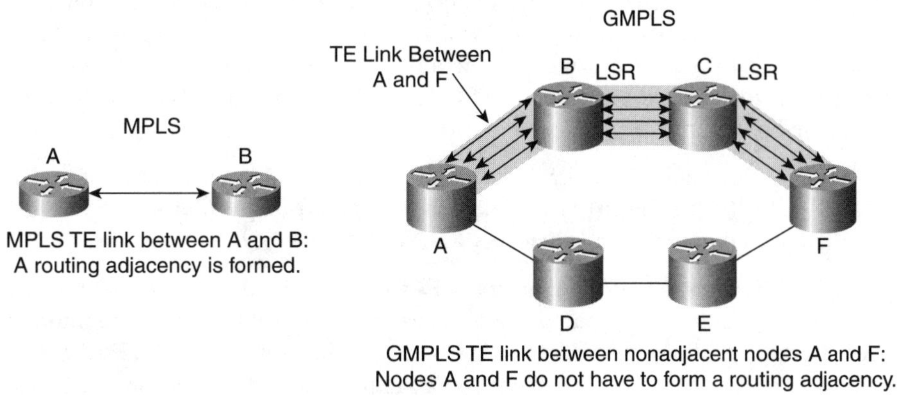

As shown in Figure 8-1, a GMPLS TE link extends beyond two adjacent nodes and can include multiple parallel component links. The end nodes of the link do not have to be part of a routing adjacency. In the context of MPLS, the link is between two adjacent nodes A and B and forms a routing adjacency using a routing protocol, say OSPF. In the GMPLS context, the link traverses multiple nodes and the two label switch routers (LSRs) B and C. A and F do not have to establish a routing adjacency.

The GMPLS enhancements to routing include the following:

- LSP hierarchy—routing
- Unnumbered links
- Link bundling
- Link protection types
- Shared link group information
- Interface switching capability descriptor

The next sections examine each of these enhancements to routing introduced by GMPLS.

LSP Hierarchy—Routing

The difference between the traditional fiber networks and WDM networks is that WDM introduces a significant increase in the number of paths between two endpoints, mainly because it introduces hundreds of wavelengths on each fiber. Couple that with the possibility of tens and hundreds of fibers between two optical switches, and the number of paths could become challenging to traditional routing protocols if every path (LSP) is considered a separate link in interior routing protocols such as OSPF and IS-IS.

LSP hierarchy can address this issue by allowing LSPs to be aggregated inside other LSPs. There is a natural order for this aggregation that is based on the multiplexing capability of the LSP types. With GMPLS, LSPs start and end on devices of the same kind, such as routers, TDM switches, WDM switches, and fiber switches. An LSP that starts and ends on a packet-switch-capable (PSC) interface can be nested with other LSPs into an LSP of type TDM that starts and ends on a TDM interface, which can be nested in LSC-LSPs that start and end on an LSC interface, which could be nested in FSC-LSPs that start and end on FSC interfaces. This is illustrated in Figure 8-2.

When an LSR establishes an LSP, it can advertise the LSP in its instance of routing protocol (OSPF or IS-IS) as a TE link. This link is called a *forwarding adjacency (FA)*. The LSP itself is referred to as the forwarding adjacency LSP, or FA-LSP.

IS-IS/OSPF floods the information about FAs just as it floods the information about any other links. As a result of this flooding, an LSR has in its TE link-state database information about not just basic TE links, but FAs as well. Figure 8-2 shows how GMPLS FA-LSP can be carried within other FA-LSPs. The different FA-LSPs introduced in this figure are FA-LSCs, FA-TDMs, and FA-PSCs.

Figure 8-2 *GMPLS LSP Hierarchy*

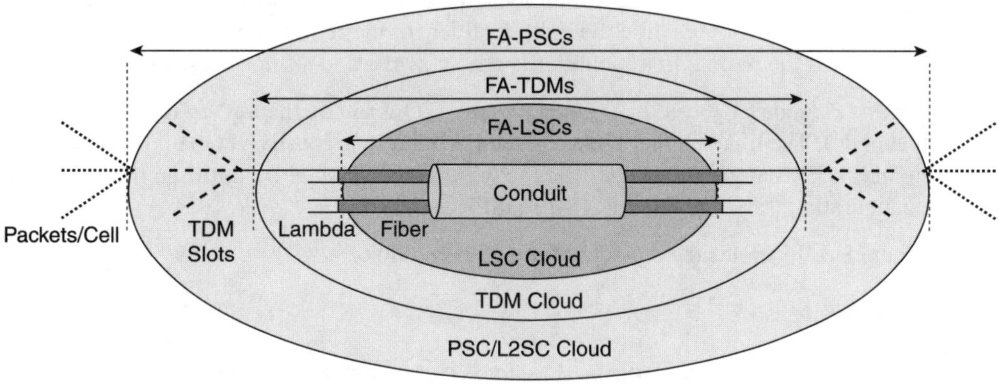

Figure 8-2 shows the following:

1 FA-LSCs are formed by nodes that sit at the boundary of a lambda cloud and a fiber cloud. The FA-LSCs get advertised in the routing protocols and are available to be used as any other TE links.

2 Nodes that sit at the boundary of a TDM cloud and a lambda cloud form FA-TDMs. The FA-TDMs get advertised as TE links.

3 Nodes that sit at the boundary of a PSC/L2SC and TDM cloud form FA-PSCs or FA-L2SCs that get advertised as TE links.

4 Low-order packet LSPs can be combined and tunneled inside higher-order FA-PSCs. In the same manner, low-order FA-PSCs can be combined and tunneled inside higher-order FA-TDMs, which can be combined and tunneled inside higher-order FA-LSCs.

5 FAs (links) are either numbered or unnumbered and can be bundled according to the GMPLS bundling procedures.

Unnumbered Links

As in an IP network, the nodes in an optical network have to be addressed and referenced. Addressing these nodes helps identify not only the nodes but also the components—that is, the links of each of these nodes. Addressing allows signaling protocols such as RSVP to establish optical paths across the OXCs.

In normal routing, each link in the network can be identified via its own subnet. This has proven to be challenging even in packet networks because it requires the assignment and management of many small subnets. In optical networks, in which the number of links can increase dramatically, IP address assignment proves much more challenging because a fiber can carry hundreds of wavelengths. Thus, the concept of unnumbered links should be quite useful.

An *unnumbered link* is a point-to-point link that is referenced using a link identifier. The link identifier is a unique, nonzero, 32-bit local identifier. The identifier for the local node is called the *local link identifier,* while the link identifier for the remote node is called the *remote link identifier.* If the remote link identifier is not known, a 0 identifier is used instead.

A network node can be addressed via a router ID (normally the highest or lowest IP address on that node). The links on that node can then be identified locally via the tuple (router ID, link number). Exchanging the identifiers may be accomplished by multiple methods, including configuration, LMP, RSVP-TE, IS-IS/OSPF, and so on.

Figure 8-3 illustrates the concept of unnumbered links.

Figure 8-3 *Unnumbered Links*

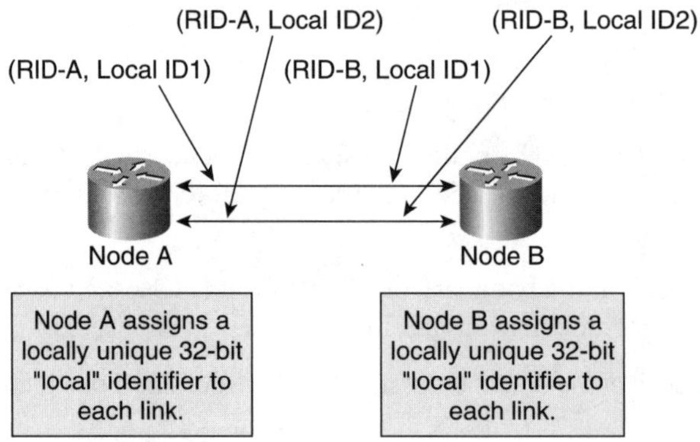

Figure 8-3 shows how node A identifies each link with a tuple formed with its router ID RID-A and the local link identifier.

Current signaling used by MPLS TE doesn't provide support for unnumbered links because the current signaling doesn't provide a way to indicate an unnumbered link in its EXPLICIT_ROUTE object (ERO) and RECORD_ROUTE object (RRO). Extensions to RSVP-TE define an optional object called LSP_TUNNEL_INTERFACE_ID that could be used in RSVP PATH or Reservation (RESV) messages. The LSP_TUNNEL_INTERFACE_ID object is an LSR router ID and a 32-bit interface ID tuple. Also, subobjects of the ERO and RRO are defined for the support of unnumbered links.

Link Bundling

Link bundling allows multiple TE links to be bundled into one bigger TE link. The subset links are called *component links,* and the group of links is called a *bundled link.*

On a bundled link, a combination of <(bundled) link identifier, component link identifier, label> is sufficient to unambiguously identify the appropriate resources used by an LSP.

Link bundling improves routing by reducing the number of links and associated attributes that are flooded into routing protocols such as OSPF and IS-IS. Link bundling allows multiple parallel links of similar characteristics to be aggregated and flooded as a bundled link. Figure 8-4 shows this concept.

Figure 8-4 *Link Bundling*

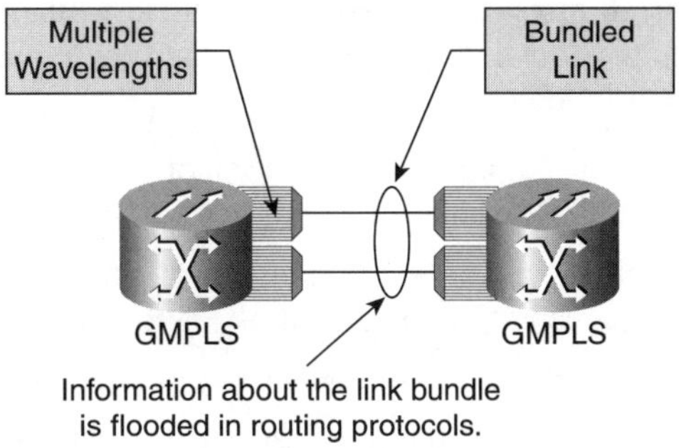

All component links in a bundle must:

- Begin and end on the same pair of LSRs
- Have the same link type, such as point-to-point or multiaccess
- Have the same TE metric
- Have the same set of resource classes at each end of the links

A bundled link is considered alive if one of its component links is alive. Determining the liveliness of the component links can be done via routing protocols, LMP, or L1 or L2 information. Once a bundled link is considered alive, the information about the bundled link is flooded as a TE link.

WARNING The benefits of link bundling in reducing the number of flooded links come at the expense of loss of information. Link bundling involves the aggregation of the component links, and in the process of summarizing the attributes of several links into a bundled link, information is lost. Remember that the information that is flooded in the routing protocols is information about the bundled link itself and *not* information about the component links. As an example, when multiple parallel SONET links are summarized, information about the total reservable bandwidth of the component links is advertised, but information about the bandwidth and time slots of each link is lost.

While the link-state protocols carry a single bundled link, signaling requires that individual component links be identified. Because the ERO does not carry information about the component links, the component link selection becomes a local matter between the LSR bundle neighbors. LMP offers a way to identify individual component links. (LMP is described later in the chapter, in the section "Link Management Protocol.")

Link Protection Types

GMPLS introduces the concept of a *link protection type,* which indicates the protection capabilities that exist for a link. Path computation algorithms use this information to establish links with the appropriate protection characteristics. This information is organized in a hierarchy where typically the minimum acceptable protection is specified at path instantiation and a path selection technique is used to find a path that satisfies at least the minimum acceptable protection. The different link protection types are as follows:

- **Extra Traffic**—This type of link protects another link or links. In case of failure of the protected links, all LSPs on this link are lost.
- **Unprotected**—This type of link is simply not protected by any other link. If the unprotected link fails, all LSPs on the link are lost.
- **Shared**—This type of link is protected by one or more disjoint links of type Extra Traffic.
- **Dedicated 1:1**—This type of link is protected by a disjoint link of type Extra Traffic.
- **Dedicated 1+1**—This type of link is protected by a disjoint link of type Extra Traffic. However, the protecting link is not advertised in the link-state database and therefore is not used by any routing LSPs.
- **Enhanced**—This type of link indicates that a protection scheme that is more reliable than Dedicated 1+1 should be used—for example, four-fiber BLSR.

Figure 8-5 shows the different protection types.

Figure 8-5 *Link Protection Types*

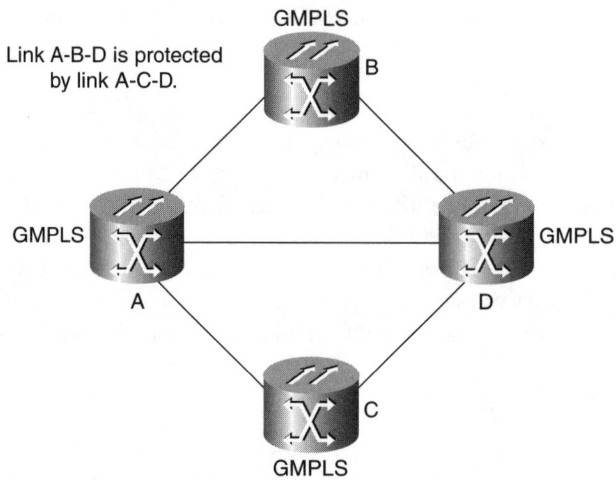

Link A-B-D is protected by link A-C-D. Link A-C-D is of type Extra Shared. The following protection scenarios can occur:

- **Link A-B-D is 1+1 protected**—Link A-C-D protects link A-B-D. Link A-C-D is not advertised and hence does not carry any LSPs unless link A-B-D fails.

- **Link A-B-D is 1:1 protected**—Link A-C-D protects link A-B-D. Link A-C-D is advertised and can carry LSPs, but it gets preempted to protect link A-B-D if link A-B-D fails.

Shared Risk Link Group Information

A set of links may constitute a *shared risk link group (SRLG)* if they share a resource whose failure may affect all links in the set. Multiple fibers in the same conduit, for example, could constitute an SRLG because a conduit cut may affect all the fibers. The same applies to multiple lambdas in a fiber that can all be affected if a fiber cut occurs. The SRLG is an optional 32-bit number that is unique within an IGP domain. A link might belong to multiple SRLGs. The SRLG of an LSP is the union of the SRLGs of the links in the LSP. The SRLG information is used to make sure that diversely routed LSPs do not have a common SRLG—that is, they do not share the same risks of failure. Figure 8-6 illustrates the concept of an SRLG.

Figure 8-6 *Shared Risk Link Group*

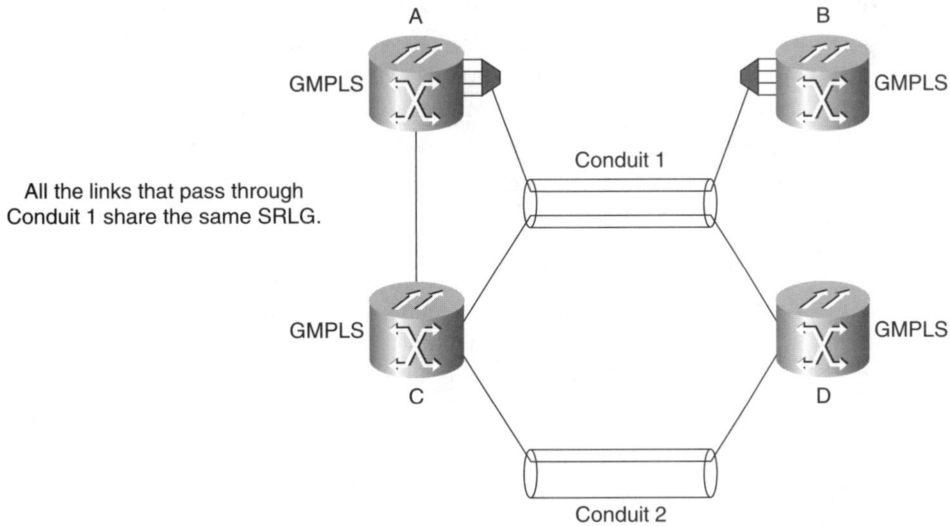

Figure 8-6 shows that all links that pass through conduit 1 share the same SRLG. The same is true for all links that pass through conduit 2. If the SRLG option is used, two LSPs that need to be diversely routed between node A and node D cannot both pass through conduit 1 or conduit 2, because they would have the same SRLGs in common.

Interface Switching Capability Descriptor

In the context of GMPLS, a link is connected to a node via an interface. An interface on the same node and on either side of the link may have multiple switching capabilities. The interface switching capability descriptor is used to handle interfaces that support multiple switching capabilities, for interfaces that have Max LSP Bandwidth values that differ by priority level (P), and for interfaces that support discrete bandwidth. A fiber interface, for example, that is connected to a node can carry multiple lambdas, and each lambda can be terminated. If the lambda is carrying packets, packet-switching can be performed. If the lambda is carrying a TDM circuit, the TDM circuit is switched. If the lambda is not terminated at the node, the lambda itself can be lambda switched. To support such interfaces, a link-state advertisement would carry a list of interface switching descriptors.

You saw in the "GMPLS Interfaces" section that GMPLS defines five types of interfaces: PSC, L2SC, TDM, LSC, and FSC. The following list describes the interface descriptors associated with these types of interfaces:

- For the PSC interfaces, various levels of PSC from 1 through 4 exist to establish a hierarchy of LSPs tunneled within LSPs, with PSC 1 being the highest order.
- For interfaces of type PSC1 through 4, TDM, and LSC, the interface descriptor carries additional information in the following manner:
  - For PSC interfaces, the additional information includes Maximum (Max) LSP Bandwidth, Minimum (Min) LSP Bandwidth, and interface MTU.
  - For TDM-capable interfaces, the additional information includes Maximum LSP Bandwidth, information on whether the interface supports standard or arbitrary SONET/SDH, and Minimum LSP Bandwidth.
  - For LSC interfaces, the additional information includes Reservable Bandwidth per priority, which specifies the bandwidth of an LSP that can be supported by the interface at a given priority number.

Determining the Link Capability

The link capability is determined based on the tuple <interface switching capability, label>. Carrying label information on a given TE link depends on the interface switching capability at both ends of the link and is determined as follows:

- [PSC, PSC]—The label is carried in the "shim" header (RFC 3032, *MPLS Label Stack Encoding*).
- [TDM, TDM]—The label represents a TDM time slot.
- [LSC, LSC]—The label represents a port on an OXC.
- [PSC, TDM]—The label represents a TDM time slot.
- [TDM, LSC]—The label represents a port.

Interface Switching Capability Descriptor Examples

The following are examples of interface switching capability descriptors.

Fast Ethernet 100-Mbps Ethernet packet interface on an LSR:

- Interface switching capability descriptor:
  - Interface Switching Capability = PSC-1
  - Encoding = Ethernet 802.3

- — Max LSP Bandwidth[P] = 100 Mbps for all P (where P indicates the LSP priority level; a priority of 7, for example, gives the LSP high priority)

The following is how the interface descriptor is represented for an OC-192 SONET interface on a digital cross-connect with Standard SONET.

Assuming that it is possible to establish the following connections, VT-1.5, STS-1, STS-3c, STS-12c, STS-48c, STS-192c, the interface switching capability descriptor of that interface can be advertised as follows:

- Interface Switching Capability = TDM [Standard SONET]
- Encoding = SONET ANSI T1.105
- Min LSP Bandwidth = VT1.5
- Max LSP Bandwidth[p] = STS192 for all p (where p refers to LSP priority)

Enhancements to Signaling

GMPLS enhances the traditional MPLS control plane to support additional different classes of interfaces, such as TDM, LSC, and FSC. The support of these interfaces requires some changes to signaling, such as the following:

- LSP hierarchy—signaling
- Enhancements to labels
- Bandwidth encoding
- Bidirectional LSPs
- Notification of label error
- Explicit label control
- Protection information
- Administrative status information
- Separation of control and data channels
- Notify messages

The following sections describe the different enhancements to signaling introduced by GMPLS.

LSP Hierarchy—Signaling

As already explained in the "LSP Hierarchy—Routing" section, GMPLS supports the concept of hierarchical LSPs, which allows multiple LSPs to be nested; that is, it allows newly initiated LSPs to be aggregated within existing LSPs. The newly initiated LSPs are tunneled inside an existing higher-order LSP, which becomes a link along the path of the new LSP. This dramatically enhances network scalability and manageability because it minimizes the number of elements that are flooded and advertised within the network. This section explains the signaling aspect of the LSP hierarchy.

To give an example of how GMPLS signaling uses the LSP hierarchy, assume that a certain router requests bandwidth to be allocated along a network consisting of data switches, SONET cross-connects, WDM-capable switches, and fiber switches.

The request from the edge router to establish a PSC LSP with a certain bandwidth could trigger the establishment of multiple higher-order LSPs that get initiated by other switches along the path. Lower-order LSPs (the new LSPs) get nested inside the higher-order LSPs that already exist or that get triggered based on the edge router's request.

Figure 8-7 shows the establishment of a series of LSPs along a path that consists of routers (R0, R1, R8, and R9), SONET ADMs (S2 and S7), WDM Optical Electrical Optical (OEO) switches (W3 and W6), and fiber switches (F4 and F5). A PATH request, path 0, needed for the formation of LSP0 between R0 and R9, is sent from R0 to R1. At router R1, this triggers the initiation of LSP1 between R1 and R8. LSP1 is nested inside LSP0. The PATH messages—path1, path2, and path3—continue to propagate, and the LSPs keep getting created until the final establishment of LSP0 between R0 and R9.

Figure 8-7 *Initiation of New Nested LSPs*

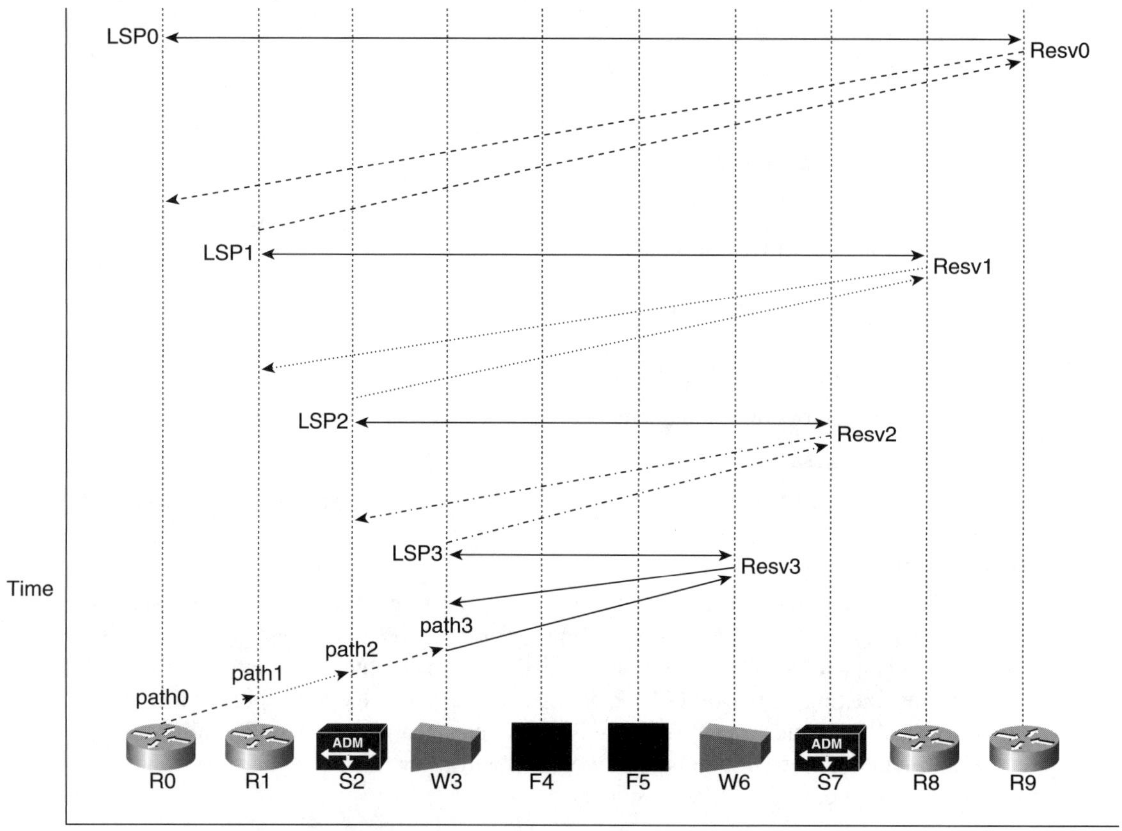

An LSP is established when the path message has completed its path inside higher-level LSPs and a RESV message is received. Note in Figure 8-8 how LSP3, the higher-level LSP, gets established first, then LSP2 gets established inside LSP3, then LSP 1 inside LSP2, and LSP 0 inside LSP1.

Figure 8-8 *Nested LSPs*

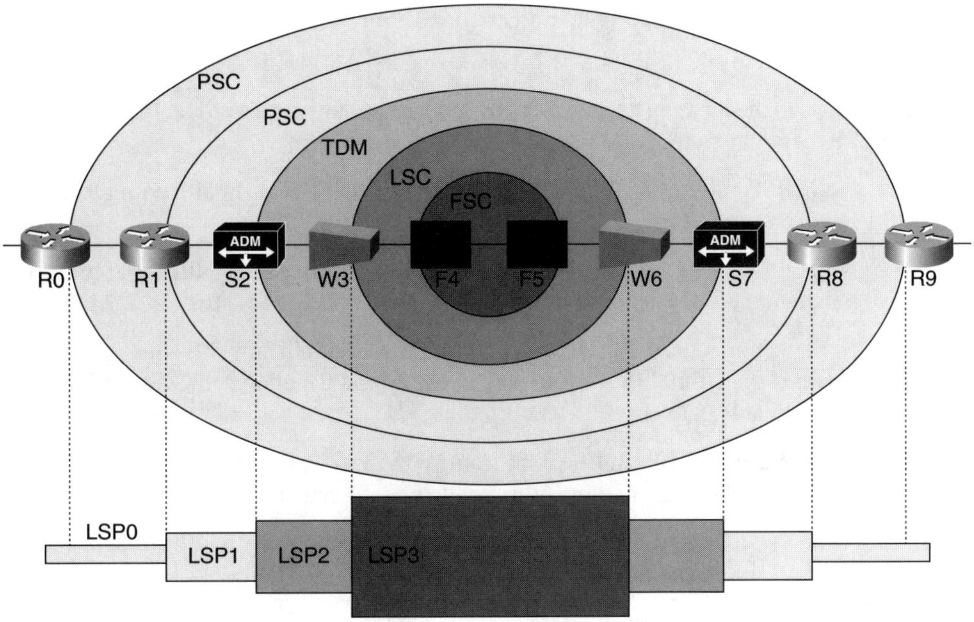

Now assume that a carrier is offering an Ethernet packet transport service between two service providers—ISP1 and ISP2—with an SLA set to 200 Mbps. For simplicity, assume that the carrier's end-to-end network is formed via routers (R0, R1, R8, and R9), SONET ADMs (S2 and S7), WDM OEO switches (W3 and W6), and fiber switches (F4 and F5). Also, for the sake of simplicity, the GE service for the carrier is assumed to be point-to-point between R0 and R9, meaning that all traffic that comes in on the GE links of R0 comes out on the GE links of R9. Physical connectivity is done in the following way:

- **R0-R1 and R8-R9**—Ethernet GE (1 Gbps) link
- **R1-S2 and R8-S7**—OC48c (2.4 Gbps) packet over SONET (PoS) link
- **S2-W3 and S7-W6**—OC192 (9.6 Gbps) TDM link
- **W3-F4 and W6-F5**—16 OC192 lambdas
- **F4-F5**—16 fibers, carrying 16 OC192 lambdas each

The following illustrates the process of LSP creation on all the boxes between ISP1 and ISP2:

- LSP0 between R0 and R9 as a 200-Mbps connection
- LSP1 between R1 and R8 as an OC48c connection
- LSP2 between S2 and S7 as an OC192 connection
- LSP3 between W3 and W6 as a lambda connection
- LSP4 between P4 and P5 as a fiber connection

LSP0 is nested inside LSP1, LSP1 is nested inside LSP2, and LSP2 is nested inside LSP3.

In addition to the creation of the LSPs, the nodes announce the residual bandwidth available in the LSP hierarchy in the following manner:

1. Node R0 announces a PSC link from R0 to R9 with bandwidth equal to the difference between the GE link and 200 Mbps—that is, 800 Mbps.
2. Node R1 announces a PSC link from R1 to R8 with bandwidth equal to the difference between the OC48c capacity (2.4 Gbps) and 200 Mbps—that is, 2.2 Gbps.
3. Node S2 announces a TDM link from S2 to S7 with bandwidth equal to the difference between the OC192 (STS-192) link capacity and the allocated OC48 (STS-48) time slots—that is, STS-144.
4. Node W3 announces an LSC link from W3 to W6 with bandwidth equal to the difference between 16 lambdas and the allocated lambda—that is, 15 lambdas.
5. Node P4 announces an FSC link from P4 to P5 with bandwidth equal to the difference between 16 fibers and the allocated fiber—that is, 15 fibers.

As part of enhancements to signaling, GMPLS introduces enhancements to the MPLS label itself, as described next.

Enhancements to Labels

GMPLS introduces new label concepts to accommodate the specific requirements of the optical space. The new concepts include the generalized label, the label set, and the suggested label.

The Generalized Label

To accommodate the scope of GMPLS that includes non-packet/cell interfaces, several new forms of labels are required, which are called *generalized labels.* A generalized label extends the traditional label by allowing the label to identify time slots, wavelengths, or space-division multiplexed positions. Examples are label representation of a fiber in a bundle, a waveband within a fiber, a wavelength in a waveband, and a set of time slots within a wavelength, as well as the traditional MPLS label. The generalized label has enough information to allow the receiving node to program a cross-connect regardless of the type of the cross-connect. As you have already seen in Chapter 7, "MPLS Controlling Optical Switches," the label is purely a signaling construct used to give information about how interfaces are cross-connected and is not part of the forwarding plane.

An example of a SONET/SDH label format is shown in Figure 8-9. This is an extension of the (K, L, M) numbering scheme defined in ITU-T Recommendation G.707, "Network Node Interface for the Synchronous Digital Hierarchy" (October 2000). The S, U, K, L, and M fields help identify the signals in the SONET/SDH multiplex. Each letter indicates a possible branch number starting at the parent node in the SONET/SDH multiplex structure.

Figure 8-9 *SONET/SDH Label Format*

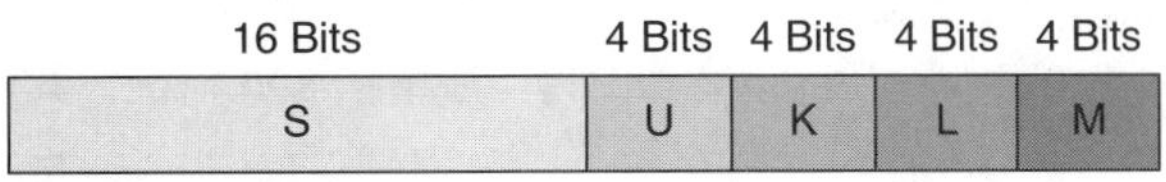

A generalized label request is used to communicate the characteristics required to support the LSP being requested. The information carried in the generalized label request includes the following:

- **LSP Encoding Type**—An 8-bit field that indicates the LSP encoding types, such as packet, Ethernet, PDH, SDH, SONET, Digital Wrapper (DW), lambda, fiber, and Fiber Channel.

  When a generalized label request is made, the request carries an LSP encoding type parameter that indicates the type of the LSP, such as SONET, SDH, Gigabit Ethernet, lambda, fiber, and so on. The lambda encoding type, for example, refers to an LSP that encompasses a whole wavelength. The fiber encoding type refers to an LSP that encompasses a whole fiber port. The encoding type represents the type of the LSP and not the nature of the links the LSP traverses. A link may support a set of encoding formats where the link can carry and switch a signal of one or more of these encoding formats depending on the link's resource availability and capacity.

- **Switching Type**—An 8-bit field that indicates the type of switching that should be performed on a particular link. This field is needed for links that advertise more than one type of switching capability, such as PSC, L2SC, TDM, LSC, and FSC.

- **Generalized Payload Identifier (G-PID)**—A 16-bit field used by the nodes at the endpoint of the LSP to identify the payload carried by the LSP. Examples of the PID are standard Ethertype values for packet and Ethernet LSPs. Other values include payload types such as SONET, SDH, Digital Wrapper (DW), STS, POS, ATM mapping, and so on.

A generalized label carries only a single level of labels—that is, the label is nonhierarchical. When multiple levels of labels are required, each LSP must be established separately, as discussed in the previous section "LSP Hierarchy—Signaling."

Waveband Switching Support

A *waveband* represents a set of contiguous wavelengths that can be switched together to a new waveband. For optimization reasons, it may be desirable for an OXC to optically switch multiple wavelengths as a unit. This may reduce the distortion on the individual wavelengths and allow tighter separation of the individual wavelengths. The waveband label is defined to support this special case.

Waveband switching uses the same format as the generalized label. Figure 8-10 shows the format of the generalized label in the context of waveband switching.

Figure 8-10 *Generalized Label—Waveband Switching*

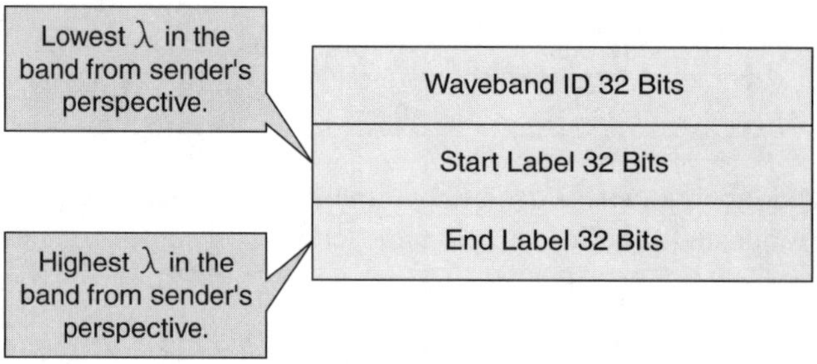

The Label Set

The *label set* is used to restrict the label ranges that may be used for a particular LSP between two peers. The receiver of a label set must restrict its choice to one label range that is in the label set. The label set is useful in the optical domain because of the restrictions on how optical equipment allocates wavelengths and handles wavelength conversion, which restricts the use of labels that are bound to these wavelengths. Reasons for using the label set include the following:

- The end equipment can transmit and receive only on a small, specific set of wavelengths/bands.
- There is a sequence of interfaces that cannot support wavelength conversion and that requires the same wavelength to be used end-to-end over a sequence of hops or an entire path.
- For operators, it is desirable to limit the amount of wavelength conversion being performed to reduce the distortion of the optical signals.
- The two ends of a link support different sets of wavelengths.

The use of a label set is optional, and if it is not present, it is assumed that all labels can be used.

The Suggested Label

GMPLS allows an upstream node to suggest a label to the downstream (one hop away) node for different optimization purposes that are specific to optical networks. The downstream node may override the suggested label at the expense of higher LSP setup times and perhaps suboptimal allocation of network resources. A typical example is when an optical switch configures its own label to adjust its mirrors and save valuable time before the downstream switch allocates the label. Other examples involve any activity where there is latency in configuring the switching fabric.

Early configuration can reduce setup latency and may be important for restoration purposes where alternate LSPs may need to be rapidly established as a result of network failures.

Bandwidth Encoding

GMPLS LSPs support packet or nonpacket LSPs. For nonpacket LSPs, it is useful to list the discrete bandwidth value of the LSP. Bandwidth encoding values include values for DS0 to OC768, E1 to STM-256, 10/100/1000/10,000-Mbps Ethernet, and 133- to 1062-Mbps Fiber Channel. The bandwidth encodings are carried in protocol-specific (RSVP-TE, CR-LDP) objects. Examples of RSVP-TE are the SENDER_TEMPLATE and FLOW_SPEC objects.

Bidirectional LSPs

Many optical service providers consider bidirectional optical LSPs a requirement, because many of the underlying constructs for SONET/SDH networks are inherently bidirectional. It is assumed that bidirectional LSPs have the same TE requirements (including fate sharing, protection, and restoration) and resource requirements (such as latency and jitter) in each direction.

The traditional MPLS LSP establishment is unidirectional. Establishing a bidirectional LSP requires establishing two unidirectional LSPs, which has many disadvantages:

- The latency to establish the bidirectional LSP is equal to one round-trip signaling time plus one initiator-terminator signaling transit delay. This extends the setup latency for successful LSP establishment and extends the worst-case latency for discovering an unsuccessful LSP. These delays are particularly significant for LSPs that are established for restoration purposes.
- The control overhead of two unidirectional LSPs is twice that of one bidirectional LSP, because separate control messages must be generated for each unidirectional LSP.
- Because the resources are established in separate segments, route selection gets complicated. Also, if the resources needed to establish the LSP are not available, one unidirectional LSP gets established, but the other doesn't. This decreases the overall probability of successful establishment of the bidirectional connection.

- SONET equipment in particular relies on hop-by-hop paths for protection switching. SONET/SDH transmits control information in-band. This requires connections to be paired, meaning that bidirectional LSP setup is highly desirable. Therefore, GMPLS supports additional methods that allow bidirectional LSP setup, to reduce session establishment overhead.

Notification of Label Error

Some situations in traditional MPLS and GMPLS result in an error message containing an "Unacceptable label value" indication. When these situations occur, it is useful if the node that is generating the error message indicates which labels are acceptable. To cover these situations, GMPLS introduces the ability to convey such information via an acceptable label set. An acceptable label set is carried in appropriate protocol-specific error messages.

The format of an acceptable label set is identical to a label set, as described earlier in this chapter in the section "The Label Set."

Explicit Label Control

As discussed in Chapter 7, with RSVP-TE, the interfaces used by an LSP may be controlled by an explicit route via the ERO or ERO hop. This allows the LSP to control which nodes/interfaces it goes in and out on. The problem is that the ERO and ERO hop do not support explicit label subobjects, which means that they cannot support the granularity needed by optical networks. For example, in networks that are not packet-based, LSPs sometimes need to be spliced together. This means that the tail end of an LSP needs to be spliced with the head end of another LSP. GMPLS introduces the ERO subobject/ERO hop to allow finer granularity for explicit routes.

Protection Information

GMPLS uses a new object type length value (TLV) field to carry LSP protection information. The use of this information is optional. Protection information indicates the LSP's link protection type. When a protection type is indicated, the connection request is processed only if the desired protection type can be honored. A link's protection capabilities may be advertised in routing.

Protection information also indicates whether the LSP is a primary or secondary LSP. A secondary LSP is a backup to a primary LSP. The resources of a secondary LSP are not used until the primary LSP fails. The resources allocated for a secondary LSP may be used by other LSPs until the primary LSP fails over to the secondary LSP. At that point, any set of LSPs that are using the resources for the secondary LSP must be preempted.

Administrative Status Information

GMPLS introduces a new object/TLV for administrative status information. The use of this information is optional. The information can be used in two ways:

- To indicate the LSP's administrative state, such as "Administratively down," "testing," or "deletion in progress." The nodes can use this information to allow local decisions, such as making sure an alarm is not sent if the LSP is put in a test mode. In RSVP-TE, this object is carried in the PATH and RESV messages.
- To send a request to set the LSP's administrative state. This request is always sent to the ingress nodes that act on the request. In RSVP-TE, this object is carried in a Notify message (discussed later, in the section "Notify Messages").

Separation of Control and Data Channels

In optical networks, the control and data channels need to be separated for multiple reasons, including these:

- Multiple links can be bundled.
- Some data channels cannot carry control information.
- The integrity of a data channel does not affect the integrity of control channels.

The following two sections discuss two critical issues for the separation of data and control channels.

Interface Identification

In MPLS, a one-to-one association exists between the data and control channels (except for MPLS link bundling). In GMPLS, where such association does not exist, it is necessary to convey additional information in signaling to identify the particular data channel being controlled. GMPLS supports explicit data channel identification by providing interface identification information. GMPLS allows the use of several interface identification schemes, including IPv4 or IPv6 addresses, interface indexes, and component interfaces (established via configuration or a protocol such as LMP). In all cases, the choice of the data interface is indicated by the addresses and identifiers used by the upstream node.

Fault Handling

Two new faults must be handled when the control channel is independent of the data channel:

- **Control channel fault**—A link or other type of failure that limits the ability of neighboring nodes to pass control messages. In this situation, neighboring nodes are unable to exchange control messages for a period of time. Once communication is restored, the underlying signaling protocol must indicate that the nodes have maintained their state through the failure. The signaling protocol must also ensure that any state changes that were instantiated during the failure are synchronized between the nodes.

- **Nodal fault**—A node's control plane fails and then restarts and loses most of its state information but does not lose its data forwarding state. In this case, both upstream and downstream nodes must synchronize their state information with the restarted node. For any resynchronization to occur, the node undergoing the restart needs to preserve some information, such as its mappings of incoming labels to outgoing labels.

Notify Messages

GMPLS provides a mechanism to inform nonadjacent nodes of LSP-related failures using *Notify messages.* In optical networks, failure notification sometimes has to traverse transparent nodes to notify the nodes responsible for restoring failed connections (transparent nodes do not originate or terminate connections). This mechanism enables target nodes to be notified directly and more quickly of a network failure. The Notify message has been added to RSVP-TE. The Notify message includes the IP address of the node that needs to be notified. Other nodes in the path just pass on the message until it reaches the targeted node. The Notify message differs from the error messages Path-Error and Reservation-Error in that it can be "targeted" to a node other than the immediate upstream and downstream neighbor.

Another application of the Notify message is to notify when the control plane has failed while the data plane is still functional. GMPLS uses this mechanism to identify *degraded* links.

Inclusion of Technology-Specific Parameters

The previous sections discussed the enhancements to signaling that allow GMPLS to control the different types of packet and nonpacket networks. GMPLS also allows the inclusion of technology-specific parameters that are carried in the signaling protocol in traffic parameter–specific objects. This section looks at how this applies to SONET/SDH. A description of parameters that are specific to optical transport network (OTN) technology is not included in this book.

The SONET/SDH traffic parameters specify a set for SONET (ANSI T1.105) and a set for SDH (ITU-T G.707), such as concatenation and transparency. Other capabilities can be defined and standardized as well. These traffic parameters must be used when SONET/SDH is specified in the LSP Encoding Type field of a generalized label request, discussed earlier in the section "The Generalized Label." The SONET/SDH traffic parameters are carried in the SENDER_TSPEC and FLOWSPEC objects of RSVP-TE and in SONET/SDH TLVs in CR-LDP.

Figure 8-11 shows how the SONET/SDH traffic parameters are organized. The Signal Type indicates the type of the elementary signal of the request LSP. Several parameters can be applied on the signal to build the final requested signals. These parameters are applied using the Request Contiguous Concatenation (RCC), Number of Contiguous Components (NCC), and Transparency fields included in the traffic parameter.

Figure 8-11 *SONET/SDH Traffic Parameters*

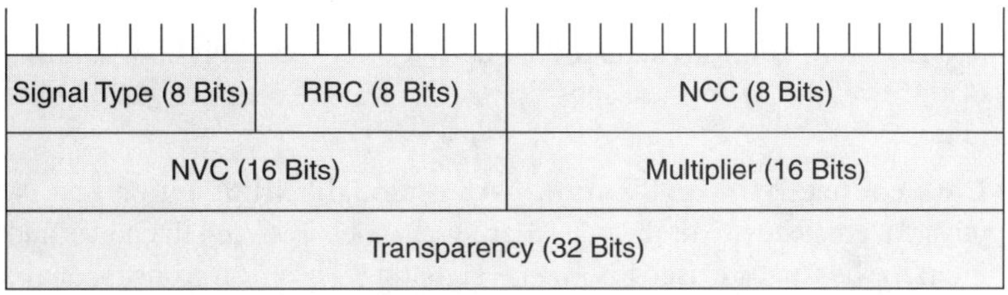

Examples of signal types for SONET/SDH include VT1.5, VT2, VT3, VT6, STS1, VC-11, VC-12, VC2, VC-3, and VC-4, plus other possible types, depending on the level of concatenation and transparency.

The RCC field, the NCC field, and the Number of Virtual Components (NVC) field are used to negotiate the type of concatenation and the number of signals that are to be concatenated. As mentioned in Chapter 2, "Metro Technologies," concatenation can be applied to signals to form larger signals. Different types of concatenation, such as contiguous or virtual, can be applied, and the information is related in the signaling protocol.

NOTE Transparency, in the context of SDH/SONET signals, refers to the overhead signals, such as the section overhead (SOH) and the line overhead (LOH) in the case of SONET. Transparency indicates which of these overhead fields needs to remain untouched when delivered to the other end of the LSP.

Link Management Protocol

Future networks may consist of optical switches, data switches that are managed by GMPLS. Thousands of fibers may connect a pair of nodes, and hundreds of wavelengths may exist on each fiber. Multiple fibers and wavelengths can be bundled to form TE links. These links need a control channel to manage routing, signaling, and link connectivity and management. LMP is a link-control protocol that runs between neighboring nodes to manage TE links.

LMP was created to address the issues of link provisioning and fault isolation to improve and scale network manageability. With GMPLS, the control channel between two adjacent nodes is no longer required to use the same physical medium as the data channels between those nodes. A control channel can run on a separate IP management network, a separate fiber, or a separate wavelength. LMP allows for the decoupling of the control channel from the component links. As such, the health of the control channel does not necessarily correlate to the health of the data links, and vice versa.

LMP is designed to provide four basic functions to a node pair:

- **Control channel management**—A core function of LMP that is used to establish and maintain control channel connectivity between neighboring nodes. This consists of lightweight Hello messages that act as a fast keepalive mechanism between the nodes.
- **Link connectivity verification**—An optional LMP function that is used to verify physical connectivity of the data-bearing channels between the nodes and to exchange the interface IDs that are used in GMPLS signaling. The error-prone manual cabling procedures make LMP link connectivity verification very useful.
- **Link property correlation**—A core function of LMP that is designed to aggregate multiple ports or component links into a TE link and to synchronize the properties of the TE link. Link properties, such as link IDs for local and remote nodes, the protection mechanism, and priority, can be exchanged via LMP using the LinkSummary message between adjacent nodes.
- **Fault management and isolation**—An optional LMP function that provides a mechanism to isolate link and channel failures in both opaque and transparent networks, irrespective of the data format. Opaque nodes are nodes where channels can be terminated for the purpose of examining the headers and data. Transparent nodes are nodes where channels pass through without termination.

LMP requires that a pair of nodes have at least one active bidirectional control channel between them. This control channel may be implemented using two unidirectional control channels that are coupled using the LMP Hello messages. LMP allows backup control channels to be defined, such as using the data-bearing channels as backup in case of failure in the primary control channels.

GMPLS Protection and Restoration Mechanisms

GMPLS introduces the necessary features in routing, signaling, and link management to support the fault management required in optical and electronic networks. Fault management requires the following capabilities:

- **Fault detection**—For optical networks, fault detection can be handled via mechanisms such as loss of light (LOL) and optical signal-to-noise ratio (OSNR) at the optical level, and bit error rate (BER), SONET/SDH Alarm Indicator Signal (AIS), or LOL at the SONET/SDH level.
- **Fault isolation**—For GMPLS, LMP can be used for fault isolation. The LMP fault-management procedure is based on sending ChannelActive and ChannelFail messages over the control channel. The ChannelActive message is used to indicate that one or more data-bearing channels are now carrying user data. The ChannelFail message is used to indicate that one or more active data channels or an entire TE link has failed.
- **Fault notification**—GMPLS uses the RSVP-TE Notify message to notify nodes of any possible failures. The Notify message can be used over the data-bearing links to indicate

a failure in the control plane, or over the control channels to indicate a failure in the data plane. The notify request object can be carried in the RSVP PATH or RESV messages and indicates the IP address of the node that should be notified when generating an error message.

GMPLS uses the following protection mechanisms:

- **1+1 protection**—The data is transmitted simultaneously over the two disjoint paths. The receiver selects the working path based on the best signal.
- **1:1 protection**—A dedicated backup path is preallocated to protect the primary path.
- **M:N protection**—M backup paths are preallocated to protect N primary paths. However, data is not replicated onto a backup path, but only transmitted in case of failure on the primary path.

  For 1:1 and M:N protection, the backup paths may be used by other LSPs. For 1+1 protection, the backup paths may not be used by other LSPs because the data is transmitted on both paths.
- **Span protection**—Intermediate nodes initiate the recovery that requires switching to an alternative path. As part of the GMPLS routing extensions, the link protection type is advertised so that span protection can be used.
- **Span restoration**—Intermediate nodes initiate the recovery that requires switching to an alternative path. The alternative path is dynamically computed.
- **Path protection**—End nodes initiate the recovery that requires switching to an alternative path. The end nodes switch to the backup path.
- **Path restoration**—End nodes initiate the recovery that requires switching to an alternative path. The backup path is dynamically calculated upon failure.

Summary of Differences Between MPLS and GMPLS

As you've learned in this chapter, GMPLS extends MPLS to support non-packet/cell interfaces. The support of the additional TDM, lambda, and fiber interfaces impacts the basic LSP properties, such as how labels are requested and communicated and the unidirectional LSP behavior, error propagation, and so on.

Table 8-1 summarizes the basic differences between MPLS and GMPLS described in this chapter.

Table 8-1 *Differences Between MPLS and GMPLS*

MPLS	GMPLS
Supports packet/cell-based interfaces only.	Supports packet/cell, TDM, lambda, and fiber.
LSPs start and end on packet/cell LSRs.	LSPs start and end on "similar type" LSRs (that is, PSC, L2SC, TDM, LSC, FSC).

continues

Table 8-1 *Differences Between MPLS and GMPLS (Continued)*

MPLS	GMPLS
Bandwidth allocation can be done in any number of units.	Bandwidth allocation can only be done in discrete units for some switching capabilities such as TDM, LSC, and FSC.
Typical large number of labels.	Fewer labels are allocated when applied to bundled links.
No restrictions on label use by upstream nodes.	An ingress or upstream node may restrict the labels that may be used by an LSP along a single hop or the whole path. This is used, for example, to restrict the number of wavelengths that can be used in the case where optical equipment provides a small number of wavelengths.
Only one label format.	Use of a specific label on a specific interface. Label formats depend on the specific interface used, such as PSC, L2SC, TDM, LSC, FSC.
Labels are used for data forwarding and are carried within the traffic.	Labels are a control plane construct only in GMPLS and are not part of the traffic.
No need for technology-specific parameters, because this is applied to packet/cell interfaces only.	Supports the inclusion of technology-specific parameters in signaling.
Data and control channels follow the same path.	Separation of control and data channels
MPLS fast-reroute.	RSVP-specific mechanism for rapid failover (Notify message)
Unidirectional LSPs.	Bidirectional LSPs enable the following: • Possible resource contention when allocating reciprocal LSPs via separate signaling sessions • Simplified failure restoration procedures • Lower setup latency • Lower number of messages required during setup
Labels cannot be suggested by upstream node.	Allow a label to be suggested by an upstream node and can be overwritten by a downstream node (to prevent delays with setting optical mirrors, for example)

Conclusion

As you have seen in this chapter, many extensions for routing, signaling, technology-specific parameters, and LMP allow the use of MPLS over non-packet/cell networks. Mechanisms such as link bundling and shared link groups are added to routing to influence the traffic trajectory and to take advantage of how the physical network topology is laid out. Signaling mechanisms such as the enhancements to the label allow the GMPLS label to be used as a control construct that indicates to the TDM/optical devices what circuits to switch and how to switch them. The introduction of LMP helps in the easy provisioning and protection of optical circuits by allowing channel link connectivity verification and fault management and isolation.

This appendix discusses the following topics:

- SONET/SDH Frame Formats
- SONET/SDH Architecture
- SONET/SDH Concatenation

APPENDIX A

SONET/SDH Basic Framing and Concatenation

This appendix explains basic SONET/SDH framing and concatenation. With the emergence of L2 metro services, SONET/SDH metro networks are being challenged to offer cost-effective and bandwidth-efficient solutions for transporting data services. The following sections describe the different elements of a SONET/SDH frame and how the elements can be combined to form bigger SONET/SDH pipes.

SONET/SDH Frame Formats

The fundamental signal in SONET is the STS-1, which operates at a rate of about 51 Mbps. The fundamental signal for SDH is STM-1, which operates at a rate of about 155 Mbps (three times the STS-1 rate). The signals are made of contiguous frames that consist of two parts: the transport overhead (TOH) contained in the header, and the payload. For synchronization purposes, the data can be allowed to shift inside the payload inside a Synchronous Payload Envelope (SPE) for SONET and inside the Virtual Container for SDH. The SPE inside the payload is referenced using a pointer. Figures A-1 and A-2 show the SONET and SDH frames.

Figure A-1 *SONET Frame Format*

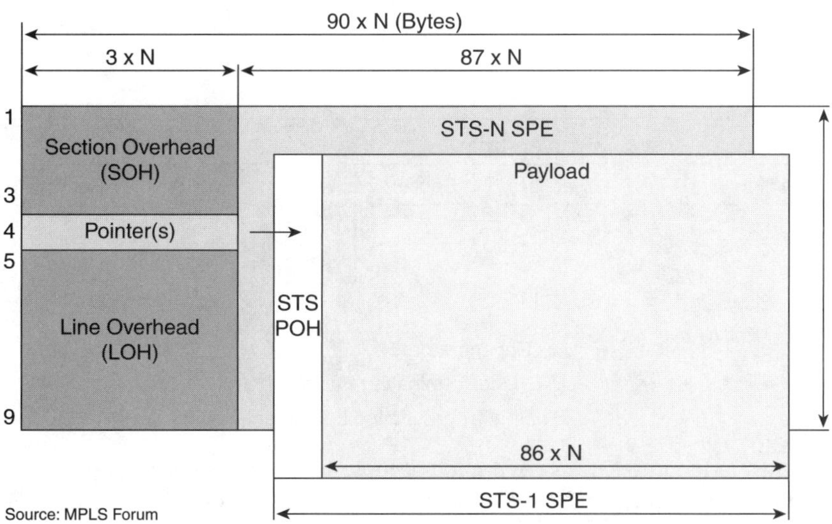

Figure A-2 *SDH Frame Format*

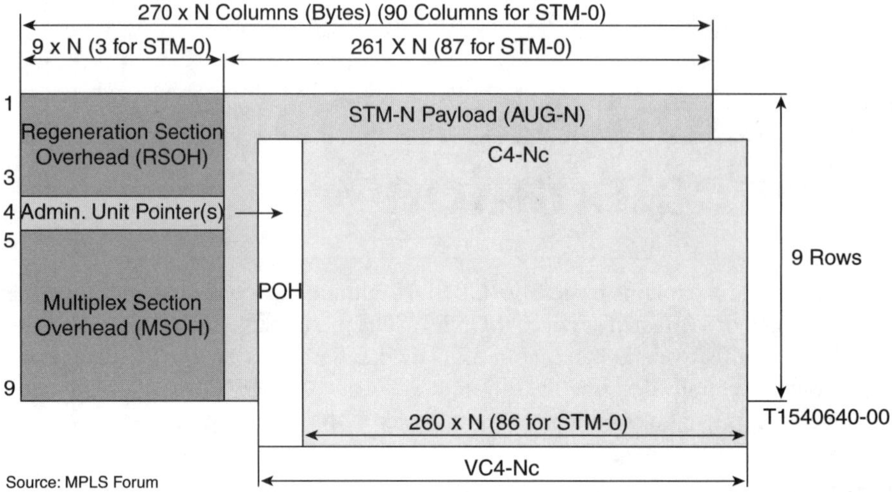

Source: MPLS Forum

SONET/SDH Architecture

The SONET/SDH architecture identifies three different layers, each of which corresponds to one level of communication between SONET/SDH equipment. The layers are as follows, starting with the lowest:

- The regenerator section, or section layer
- The multiplex section, or line layer
- The path layer

Figure A-3 shows the three SONET/SDH layers.

Figure A-3 *SONET and SDH Layers*

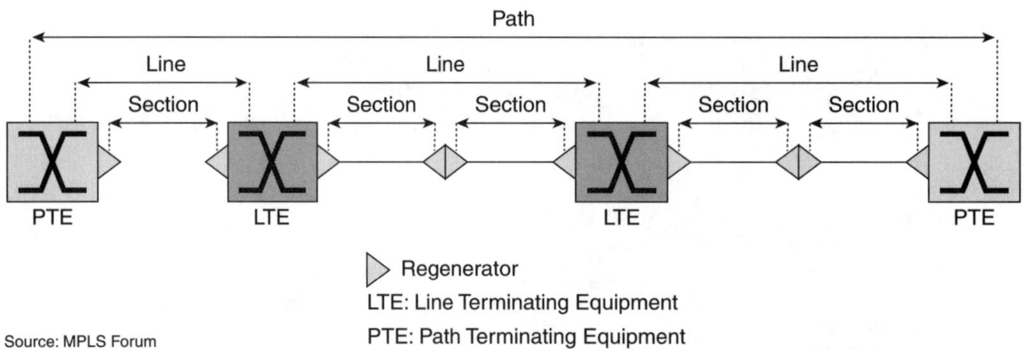

Source: MPLS Forum

As shown in Figures A-1 and A-2, each of these layers in turn has its own overhead (header). The transport overhead (TOH) of a SONET/SDH frame is mainly subdivided into two parts that contain the section overhead (SOH) and the line overhead (LOH). In addition, a pointer indicates the beginning of the SPE/Virtual Container in the payload of the overall frame. The SPE/Virtual Container itself is made up of the path overhead (POH) and a payload. This payload can be further subdivided into subelements, or a multiplex structure (signals). This multiplex structure leads to identifying time slots that contain tributary signals such as T1 (1.5 Mbps), E1 (2 Mbps), and so on. For example, a SONET STS-1 can be further divided into 7 * VT-6 (virtual tributaries), where VT-6 is equal to 6.321 Mbps. A VT-6 can be divided into 4 * VT 1.5, where a VT-1.5 is 1.544 Mbps or a T1. Figure A-4 shows the SONET multiplexing structure. Figures A-5 and A-6 show the SDH multiplexing structure. Table A-1 shows some helpful mapping between SONET and SDH.

Figure A-4 *SONET High-Order and Low-Order Multiplexing Structure*

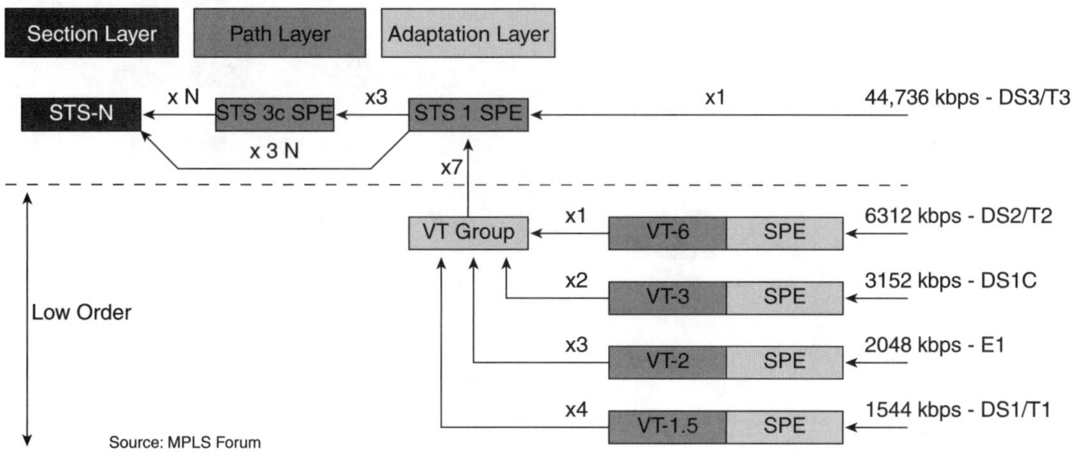

Figure A-5 *SDH High-Order Multiplexing Structure*

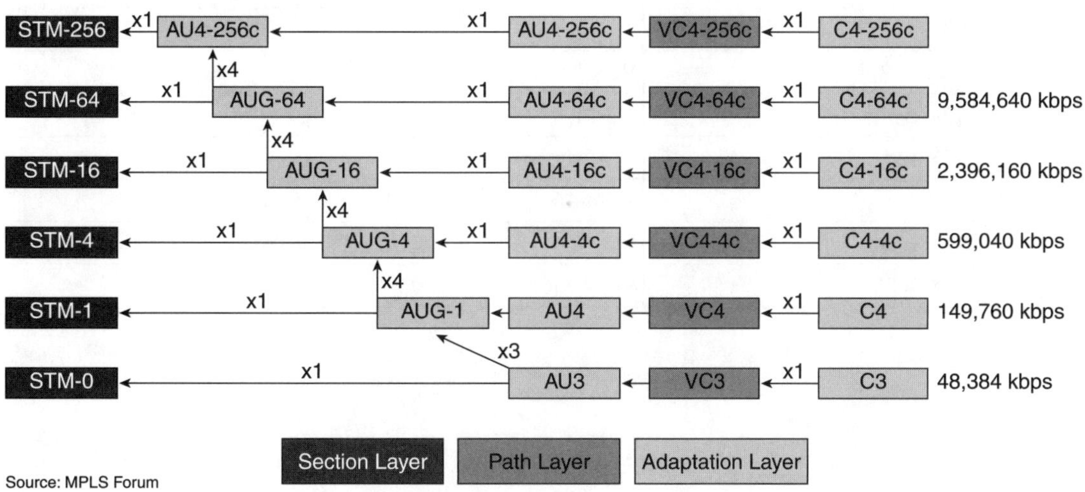

Figure A-6 *SDH Low-Order Multiplexing Structure*

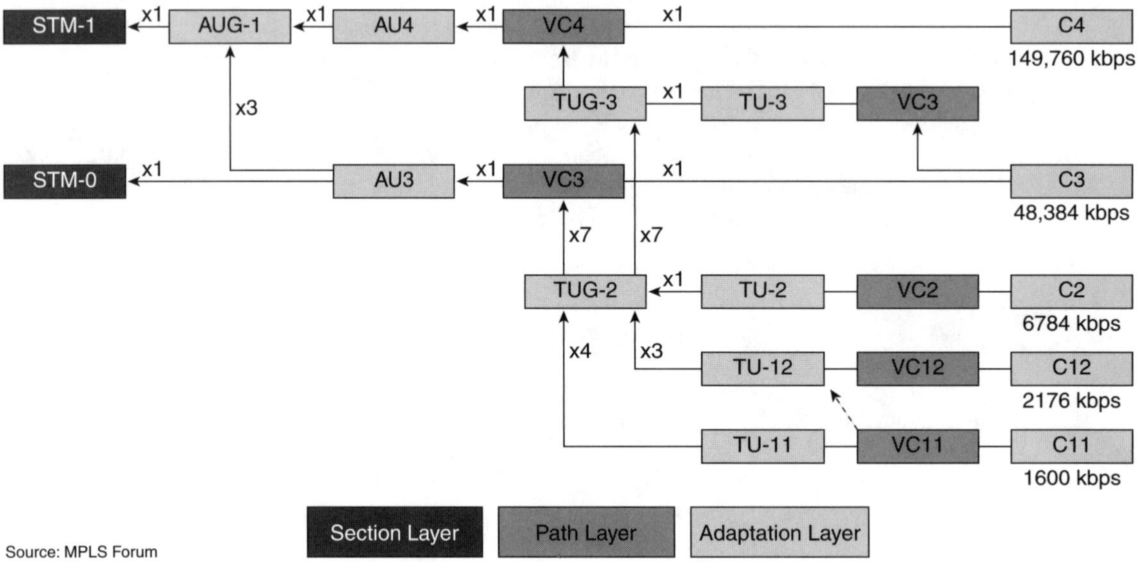

Source: MPLS Forum

Table A-1 *Helpful SONET/SDH Equivalency*

SONET	SDH	
STS-1	VC-3	STM-0
STS-3c	VC-4	STM-1
VT-6	VC-2	
VT-3		
VT-2	VC-12	
VT-1.5	VC-11	
STS-12c	VC-4-4c	STM-4
STS-48c	VC-4-16c	STM-16
STS-192c	VC-4-64c	STM-64
STS-768c	VC-4-256c	STM-256

NOTE Note that STS-3, -12, -48, -192, -768, and so on are referred to as OC-3, -12, -48, -192, and so on.

An STS-*N*/STM-*N* signal is formed from *N* STS-1/STM-1 signals via byte interleaving. The SPEs/Virtual Containers in the *N* interleaved frames are independent and float according to their own clocking. This means that an STS-3 (OC3) pipe with bandwidth of about 155 Mbps is formed from three STS-1 signals. An STS-12 (OC12) pipe with bandwidth of about 622 Mbps is formed from 12 STS-1 signals. The STS-1 signals are independent.

SONET/SDH Concatenation

To transport tributary signals in excess of the basic STS-1/STM-1 signal rates, the SPEs/Virtual Containers can be concatenated—that is, glued together. In this case, their relationship with respect to each other is fixed in time, and they act as one bonded pipe.

Different types of concatenations are defined, including *contiguous standard concatenation* and *virtual concatenation.*

Contiguous Standard Concatenation

Contiguous standard SONET concatenation allows the concatenation of *M* STS-1 signals within an STS-*N* signal, with $M <= N$ and $M = 3, 12, 48, 192, 768$, and so on in multiples of 4. The SPEs of these *M* STS-1s can be concatenated to form an STS-*M*c. The STS-*M*c notation is shorthand for describing an STS-*M* signal whose SPEs have been concatenated (c stands for concatenated). This means that an STS-12c (OC12c) is formed from the concatenation of 12 STS-1 signals, and the 12 STS-1s act as one bonded pipe. Constraints are imposed on the size of STS-*M*c signals (that is, they must be a multiple of 3) and on their starting location and interleaving.

Figure A-7 shows an example of a SONET OC192 pipe (9.6 Gbps) that is multiplexed into 192 STS-1s or into four concatenated STS-48c pipes.

One of the disadvantages of standard concatenation is the lack of flexibility in starting time slots for STS-*M*c signals and in their interleaving. This means that the provider has to deploy SONET/SDH circuits with the predefined concatenation bandwidth size and with bandwidth increments that do not match its customer needs. This leads to inefficiencies in bandwidth deployment. Virtual concatenation solves this problem.

Figure A-7 *Sample SONET Structure*

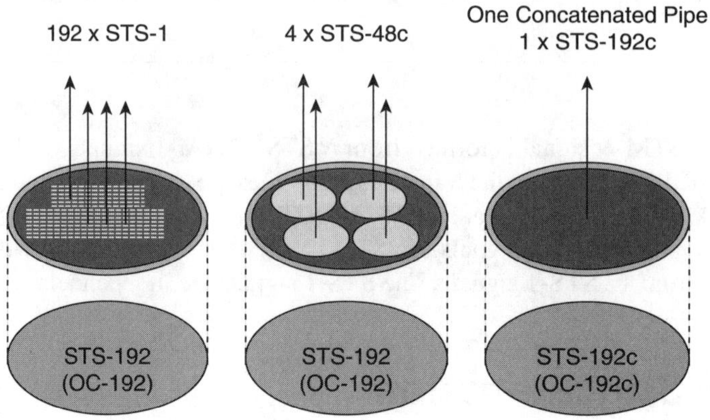

Virtual Concatenation

Virtual concatenation is a SONET/SDH end-system service approved by the committee T1 of ANSI and ITU-T. The essence of this service is to have SONET/SDH end systems "glue" together the Virtual Containers or SPEs of separate signals rather than requiring that the signals be carried through the network as a single unit. In one example of virtual concatenation, two end systems that support this feature could essentially combine two STS-1s into a virtual STS-2c for the efficient transport of 100-Mbps Ethernet traffic. If instead these two end systems were to use standard concatenation with increments of STS-1, STS-3, and STS-12, a 100-Mbps pipe would not fit into an STS-1 (51 Mbps) circuit and would have to use an STS-3c (155 Mbps) circuit, therefore wasting about 55 Mbps of bandwidth. By using a virtual-concatenated STS-2c circuit (around 100 Mbps), the operator can achieve 100 percent efficiency in transporting a 100-Mbps Ethernet pipe.

NOTE The industry has suggested the use of *arbitrary contiguous concatenation*, which is similar in nature to virtual concatenation; however, it is applied inside the SONET/SDH network rather than the SONET/SDH end systems. Virtual concatenation will emerge as the solution of choice for next-generation data over SONET/SDH network deployments.

Conclusion

This appendix has presented the basics of SONET/SDH framing and explained how the SONET/SDH technology is being adapted via the use of standard and virtual concatenation to meet the challenging needs of emerging data over SONET/SDH networks in the metro. The emergence of L2 metro services will challenge the legacy SONET/SDH network deployments and will drive the emergence of multiservice provisioning platforms (MSPPs) that will efficiently transport Ethernet, Frame Relay, ATM, and other data services over SONET/SDH.

GLOSSARY

A

add/drop multiplexer (ADM). A device installed at an intermediate point on a transmission line that enables new signals to come in and existing signals to go out. Add/drop multiplexing can be done with optical or electronic signals. The device may deal only with wavelengths, or it may convert between wavelengths and electronic TDM signals.

adjacency. A relationship formed between selected neighboring routers and end nodes for the purpose of exchanging routing information.

B

black hole. Routing term for an area of the internetwork where packets enter but do not emerge due to adverse conditions or poor system configuration within a portion of the network.

Building Local Exchange Carriers (BLECs). Service providers that offer broadband services to businesses and tenants concentrated in building offices.

C

class of service (CoS). A classification whereby different data packets that belong to a certain class receive similar quality of service.

committed burst size (CBS). A parameter associated with CIR that indicates the size up to which subscriber traffic is allowed to burst in profile and not be discarded or shaped.

committed information rate (CIR). The minimum guaranteed throughput that the network must deliver for the service under normal operating conditions.

component link. A subset of a bigger link. A channel within a SONET/SDH channelized interface is an example of a component link.

control plane. A logical plane where protocol packets get exchanged for the purpose of achieving multiple functions, such as setting up paths used for packet forwarding or for managing the nodes in the network.

customer edge (CE) device. A device such as a switch or router that resides at the customer premises. The device could be owned by the customer or the provider.

customer premises equipment (CPE). Terminating equipment, such as switches, routers, terminals, telephones, and modems, supplied by the telephone company, installed at customer sites, and connected to the telephone company network.

D

Data Packet Transport (DPT). A Media Access Control protocol that adds resiliency and protection to packet networks deployed in a ring topology.

Decoupled Transparent LAN Service (DTLS). A service that emulates a LAN over an IP/MPLS network, similar to VPLS. DTLS, however, proposes to remove any L2 switching from the provider edge devices and restrict the L2 switching to the customer edge devices.

detour LSP. An LSP that is set up to reroute the traffic in case the main LSP fails.

Diffserv. A method used to classify IP packets so that different classes receive different quality of service treatment when forwarded in the network.

E

Ethernet LAN Service (E-LAN). A multipoint-to-multipoint Ethernet service.

Ethernet over MPLS (EoMPLS). An L2 tunneling technique that allows Ethernet frames to be carried over an IP/MPLS network.

Ethernet over SONET/SDH (EOS). A technology that allows Ethernet packets to be transported over a SONET/SDH TDM network.

Ethernet Virtual Connection (EVC). A point-to-point Ethernet service.

Explicit Route Object (ERO). A field that indicates the path to be taken when traffic is forwarded.

F

fiber-switch capable (FSC) interfaces. Interfaces that can forward data based on a position of the data in the real-world physical spaces. This is typical of optical cross-connects that switch traffic on the fiber or multiple-fiber level.

Fixed Filter (FF). Reservation style that creates a distinct reservation for traffic from each sender. This style is common for applications in which traffic from each sender is likely to be concurrent and independent. The total amount of reserved bandwidth on a link for sessions using FF is the sum of the reservations for the individual senders.

Forwarding Information Base (FIB). A data structure and way of managing forwarding in which destinations and incoming labels are associated with outgoing interfaces and labels.

frame check sequence (FCS). Extra characters added to a frame for error control purposes. Used in HDLC, Frame Relay, and other data link layer protocols.

G

generalized label request. An MPLS label scheme that extends the use of the MPLS label to nonpacket networks.

Generalized Multiprotocol Label Switching (GMPLS). A generalized MPLS control plane that allows the provisioning and protection of circuits over both packet and nonpacket networks.

Generic Attribute Registration Protocol (GARP). A protocol defined by the IEEE to constrain multicast traffic in bridged Ethernet networks.

Gigabit Ethernet (GE). Standard for a high-speed Ethernet, approved by the IEEE 802.3z standards committee in 1996.

I

incumbent local exchange carrier (ILEC). Traditional telephony company.

interexchange carrier (IXC). Common carrier that provides long-distance connectivity between dialing areas serviced by a single local telephone company.

interface. In routing or transport terminology, a network connection or a port.

Interior Gateway Protocol (IGP). Internet protocol used to exchange routing information within an autonomous system. Examples of common Internet IGPs include IGRP, OSPF, and RIP.

L

L2TPv3. An L2 tunneling protocol that allows the tunneling of Ethernet packets over an L3 IP network.

label block. A block of MPLS labels exchanged between two MPLS routers.

Label Distribution Protocol (LDP). A standard protocol between MPLS-enabled routers to negotiate the labels (addresses) used to forward packets. The Cisco proprietary version of this protocol is the Tag Distribution Protocol (TDP).

label switch router (LSR). Forwards packets in an MPLS network by looking only at the fixed-length label.

label switched path (LSP). A path that MPLS packets traverse between two edge LSRs.

lambda-switch capable (LSC) interfaces. Interfaces that can forward data based on the wavelength on which it was received. This is typical of optical cross-connects that switch traffic on the wavelength level.

Layer 2-switch capable (L2SC) interfaces. Interfaces that can recognize L2 cell or frame boundaries and can forward data based on L2 headers. This is typical of interfaces on ATM switches, Frame Relay switches, and L2 Ethernet switches.

Link Aggregation Control Protocol (LACP). A protocol that allows multiple Ethernet links to be bundled in a larger pipe.

link bundling. Aggregating multiple links into a bigger pipe.

Link Management Protocol (LMP). Establishes and maintains control channel connectivity between neighbors. LMP also enables neighbor discovery, which allows neighbors to identify connected devices, obtain UNI connectivity information, and identify and verify port-level connections, network-level addresses, and corresponding operational states for every link.

LSP tunnel. A configured connection between two routers that uses MPLS to carry the packets.

M

maximum transmission unit (MTU). Maximum packet size, in bytes, that a particular interface can handle.

Media Access Control (MAC) address. Standardized data link layer address that is required for every port or device that connects to a LAN. Other devices in the network use these addresses to locate specific ports in the network and to create and update routing tables and data structures. MAC addresses are 6 bytes long and are controlled by the IEEE.

metropolitan area network (MAN). Network that spans a defined metropolitan or regional area; smaller than a WAN but larger than a LAN.

multidwelling units (MDUs). Buildings that contain multiple housing units, such as apartment complexes and university dormitories.

multiple service operator (MSO). Cable service provider that also provides other services, such as data and voice telephony.

multiplexing. Scheme that allows multiple logical signals to be transmitted simultaneously across a single physical channel.

multipoint-to-multipoint (MP2MP). An any-to-any connection between end systems.

Multiprotocol Label Switching (MPLS). Switching method that forwards IP traffic using a label. This label instructs the routers and switches in the network where to forward the packets based on pre-established IP routing information.

multitenant units (MTUs). Multitenant building offices that are recipients of broadband services by a BLEC.

N

Network Management System (NMS). System responsible for managing at least part of a network. An NMS is generally a reasonably powerful and well-equipped computer, such as an engineering workstation. NMSs communicate with agents to help keep track of network statistics and resources.

Network-to-Network Interface (NNI). A specification of the interface between a backbone system and another backbone system. For example, the specification of an optical interface that connects two optical switches in the carrier network.

Notify message. A message used by RSVP-TE to notify other nodes of certain failures.

O

OAM&P. Operations, administration, maintenance, and provisioning. Provides the facilities and personnel required to manage a network.

optical cross-connect (OXC). A network device that switches high-speed optical signals.

P

packet multiplexing. Data packets coming in from different locations and being multiplexed over the same output wire.

packet-switch capable (PSC). Systems such as IP/MPLS routers that can switch data packets.

Packet-switch capable (PSC) interfaces. Interfaces that can recognize packet boundaries and can forward data based on packet headers. This is typical of interfaces on routers and Layer 3 Ethernet switches.

packet switching. The ability to forward packets in the network based on packet headers or fixed labels.

peak information rate (PIR). Specifies the maximum rate above the CIR at which traffic is allowed into the network and may get delivered if the network is not congested.

point of local repair (PLR). The router at which a failed LSP can be locally rerouted.

point-to-point (P2P). A one-to-one connection between two end systems.

provider (P) device. Normally, a core IP/MPLS router that offers a second level of aggregation for the provider edge devices.

provider edge (PE) device. A provider-owned device that offers the first level of aggregation for the different customer edge (CE) devices.

pseudowire (PW). A representation of packet-leased line, or a virtual circuit between two nodes.

Q

Q-in-Q. An Ethernet encapsulation technique that allows Ethernet packets that already have an 802.1Q VLAN tag to be 802.1Q VLAN tagged again.

R

Record Route Object (RRO). A field that indicates the path that traffic takes when forwarded.

regional Bell operating company (RBOC). Regional telephone company formed by the breakup of AT&T.

Resilient Packet Ring (RPR). A Media Access Control standard protocol that adds resiliency and protection to packet networks deployed in a ring topology.

Resource Reservation Protocol (RSVP). Protocol that supports the reservation of resources across an IP network. Applications running on IP end systems can use RSVP to indicate to other nodes the nature (bandwidth, jitter, maximum burst, and so on) of the packet streams they want to receive.

RSVP-TE. A protocol that extends RSVP to support traffic engineering over an IP/MPLS network.

S

Shared Explicit (SE). Reservation style that allows a receiver to explicitly select a reservation for a group of senders, rather than one reservation per sender, such as in the FF style. Only a single reservation is shared between all senders listed in the particular group.

shared risk link group (SRLG). A grouping that indicates similar risk characteristics for a set of elements. A set of fibers, for example, that share the same conduit belong to the same SRLG, because if the conduit is cut, all fibers will fail.

shortest path first (SPF) algorithm. Routing algorithm that iterates on length of path to determine a shortest-path spanning tree. Commonly used in link-state routing algorithms. Sometimes called Dijkstra's algorithm.

SONET/SDH terminal multiplexer (TM). A device installed at an endpoint on a transmission line that multiplexes multiple transmission lines, such as DS1s/DS3s, into a SONET/SDH network.

spanning tree. Loop-free subset of a network topology.

Synchronous Digital Hierarchy (SDH). A standard for delivering data over optical fiber. SDH is used in Europe.

Synchronous Optical Network (SONET). A standard for delivering data over optical fiber. SONET is used in North America and parts of Asia.

Synchronous Payload Envelope (SPE). The payload-carrying portion of the STS signal in SONET. The SPE is used to transport a tributary signal across the synchronous network. In most cases, this signal is assembled at the point of entry to the synchronous network and is disassembled at the point of exit from the synchronous network. Within the synchronous network, the SPE is passed on intact between network elements on its route through the network.

T

time-division multiplexing (TDM). Technique in which information from multiple channels can be allocated bandwidth on a single wire based on preassigned time slots. Bandwidth is allocated to each channel regardless of whether the station has data to transmit.

time-division multiplexing (TDM) interfaces. Interfaces that can recognize time slots and can forward data based on the data's time slot in a repeating cycle. This is typical of interfaces on digital cross-connects, SONET ADMs, and SONET cross-connects.

Time To Live (TTL). A mechanism to prevent loops in IP networks. The TTL field gets decremented every time a packet traverses a router. When TTL reaches 0, the packet can no longer be forwarded.

traffic engineered (TE) link. A link that is set up to divert the traffic over a path different than what is calculated by Interior Gateway Protocols.

traffic engineering (TE). Techniques and processes that cause routed traffic to travel through the network on a path other than the one that would have been chosen if standard routing methods were used.

traffic engineering (TE) tunnel. A label-switched tunnel that is used for traffic engineering. Such a tunnel is set up through means other than normal L3 routing; it is used to direct traffic over a path different from the one that L3 routing could cause the tunnel to take.

traffic trunk. Physical and logical connection between two switches across which network traffic travels. A backbone is composed of a number of trunks.

Transparent LAN Service (TLS). A service that extends the LAN over the MAN and WAN.

trunk. Physical and logical connection between two switches across which network traffic travels. A backbone is composed of a number of trunks.

tunnel. A connection between two end systems that allows the encapsulation of packets within it.

U

unidirectional path switched ring (UPSR). Path-switched SONET rings that employ redundant, fiber-optic transmission facilities in a pair configuration. One fiber transmits in one direction, and the backup fiber transmits in the other. If the primary ring fails, the backup takes over.

unnumbered link. A link that does not have an IP address assigned to it.

User-to-Network Interface (UNI). A specification of the interface between an end system and a backbone system. An example is the specification of an Ethernet interface that connects a switch at the customer site and a router at the provider site.

V

virtual circuit (VC). Logical circuit created to ensure reliable communication between two network devices. A virtual circuit is defined by a VPI/VCI pair, and can be either permanent (PVC) or switched (SVC).

Virtual Container. An SDH signal that transports payloads that are smaller than an STM-0 (48,384 kbps) payload. VC is part of the SDH hierarchy.

virtual LAN (VLAN). Group of devices on one or more LANs that are configured (using management software) so that they can communicate as if they were attached to the same wire, when in fact they are located on a number of different LAN segments. Because VLANs are based on logical instead of physical connections, they are extremely flexible.

Virtual Private LAN Service (VPLS). A service that extends the notion of a switched Ethernet LAN over an IP/MPLS network.

virtual router forwarding (VRF). A VPN routing/forwarding instance. A VRF consists of an IP routing table, a derived forwarding table, a set of interfaces that use the forwarding table, and a set of rules and routing protocols that determine what goes into the forwarding table. In general, a VRF includes the routing information that defines a customer VPN site that is attached to a PE router.

Virtual Tributary. A SONET signal that transports payloads that are smaller than an STS-1 (44,736 kbps) payload. VT is part of the SONET hierarchy.

W

waveband. A set of contiguous wavelengths that can be switched together as a unit.

wavelength-division multiplexing (WDM). Optical technology whereby multiple optical wavelengths can share the same transmission fiber. The spectrum occupied by each channel must be adequately separated from the others.

Wildcard Filter (WF). Reservation style in which a single shared reservation is used for all senders to a session. The total reservation on a link remains the same regardless of the number of senders.

INDEX

D

E

F

G

H-J

L

M

N

O

P

Q-R

S

T

U

V

W-X